Energy Security in the 1980s:
Economic and Political
Perspectives

Energy Security in the 1980s: Economic and Political Perspectives

A Staff Paper by
Douglas R. Bohi and William B. Quandt

THE BROOKINGS INSTITUTION
Washington, D.C.

© 1984 by
THE BROOKINGS INSTITUTION
1775 Massachusetts Avenue, N.W., Washington, D.C. 20036

Library of Congress Catalogue Card Number 84-72355
ISBN 0-8157-1001-1

1 2 3 4 5 6 7 8 9

THE BROOKINGS INSTITUTION is an independent organization devoted to nonpartisan research, education, and publication in economics, government, foreign policy, and the social sciences generally. Its principal purposes are to aid in the development of sound public policies and to promote public understanding of issues of national importance.

The Institution was founded on December 8, 1927, to merge the activities of the Institute for Government Research, founded in 1916, the Institute of Economics, founded in 1922, and the Robert Brookings Graduate School of Economics and Government, founded in 1924.

The Board of Trustees is responsible for the general administration of the Institution, while the immediate direction of the policies, program, and staff is vested in the President, assisted by an advisory committee of the officers and staff. The by-laws of the Institution state: "It is the function of the Trustees to make possible the conduct of scientific research, and publication, under the most favorable conditions, and to safeguard the independence of the research staff in the pursuit of their studies and in the publication of the results of such studies. It is not a part of their function to determine, control, or influence the conduct of particular investigations or the conclusions reached."

The President bears final responsibility for the decision to publish a manuscript as a Brookings book. In reaching his judgment on the competence, accuracy, and objectivity of each study, the President is advised by the director of the appropriate research program and weighs the views of a panel of expert outside readers who report to him in confidence on the quality of the work. Publication of a work signifies that it is deemed a competent treatment worthy of public consideration but does not imply endorsement of conclusions or recommendations.

The Institution maintains its position of neutrality on issues of public policy in order to safeguard the intellectual freedom of the staff. Hence interpretations or conclusions in Brookings publications should be understood to be solely those of the authors and should not be attributed to the Institution, to its trustees, officers, or other staff members, or to the organizations that support its research.

Foreword

IN THE 1970s the dramatic price shocks associated with the Arab-Israeli war of October 1973 and the Iranian revolution of 1979 resulted in an "energy crisis." Long lines at gas stations vividly symbolized the problems of adjusting to temporary shortages of oil. Even though that particular crisis is over and gas lines are unlikely to reappear in the 1980s, energy security remains an important issue.

In 1980 the Brookings Institution and Resources for the Future began a systematic program of research, supported primarily by the Department of Energy, to examine the economic and political dimensions of energy security in the 1980s. This study draws extensively on that research and looks ahead to how the energy problem is likely to develop in the rest of the decade. Douglas R. Bohi, a senior fellow at Resources for the Future, and William B. Quandt, a senior fellow at Brookings, have brought together their respective economic and political expertise to address this question.

According to Bohi and Quandt, energy markets in the 1980s are expected to work more efficiently than they did in the 1970s, with less risk of large and enduring price shocks. The authors caution, however, that political developments, especially in the Middle East, could still lead to disruptions of oil supplies and a rise in prices. But once the disruptions have ended, consumers should expect prices to decline.

This study emphasizes that the energy security problem of the 1980s will differ significantly from that of the 1970s. Lessons from the past are not always a good guide to the future. The authors do not, however, suggest that the energy problem is behind us and that markets alone can be left to work their magic. Diplomacy, military preparedness, and public policy still have a role in reducing the risks of future energy crises and in dealing with the consequences of oil-supply disruptions.

The authors have benefited from the comments of their colleagues at Resources for the Future and Brookings. In particular, thanks go to Harry Broadman, Joel Darmstadter, Joy Dunkerley, Edward R. Fried, Ed A. Hewett, Hans Landsberg, Richard P. Mattione, Thomas L. McNaugher, John D. Steinbruner, and Michael Toman. Ted Moran and Lincoln Gordon reviewed the manuscript and made useful suggestions. Ruth E. Conrad served as administrative assistant to the energy and national security program and typed the manuscript; Gregg Forte edited it; and Alan G. Hoden verified its factual content.

vii

Funding for this study came largely from the U.S. Department of Energy. Additional funding was provided by the Rockefeller Foundation.

The views expressed here are those of the authors and should not be ascribed to the persons or organizations whose assistance is acknowledged, to Resources for the Future, or to the trustees, officers, or other staff members of the Brookings Institution.

BRUCE K. MAC LAURY
President

September 1984
Washington, D.C.

LOOKING back over the past decade, one can clearly see that the international energy picture has been shaped by both economic conditions and political developments and that neither dimension, by itself, can tell the whole story. To be sure, without much more than an elementary appreciation of the underlying economics of supply and demand, analysts could have predicted that worldwide trends in oil consumption and production from the 1960s could not last for long. However, two political crises in the Middle East proved to be critical in galvanizing the changes: the Arab-Israeli war of 1973 and the Iranian revolution in 1978–79. Each helped to set in motion a diverse set of economic and political responses that contributed to the restructuring of international petroleum markets.

There were special market conditions in the 1970s that allowed individual members of the Organization of Petroleum Exporting Countries, aided by these two Middle East crises, to gain control over oil company operations in their own countries and to influence market prices through decisions on output and the development of reserves. OPEC's successes created the impression of a powerful cartel composed of members who were prepared to cooperate to advance their collective interests. From the vantage point of the 1980s it is possible to see that the huge price increases in 1973 and 1979 were only in part due to OPEC's deliberate policies. They were also the result of special market conditions: consumer behavior based on expectations of rising prices, international political developments that temporarily disrupted oil markets, and public policies in consuming countries that magnified the impact of the oil shortages. OPEC in a sense was an organization riding the crest of a wave that it had done little to create.

The events of the 1970s led to changes in both economic and political conditions. There were several reactions to the increased price of oil: economic growth slowed in the industrial countries, and the demand for energy began to subside; incentives were created for more efficient use of energy and, in particular, for switching from oil to other fuels; oil exploration and development outside of OPEC were stimulated; and important changes began to emerge in the way petroleum was traded in world markets.

For the foreseeable future, political and military events are likely to be

1

the primary catalysts for sudden and dramatic shocks to energy markets. Economic conditions will determine how these shocks play out over time by shaping incentives for private actions. Yet, in both dimensions, it is necessary to understand that the energy problem likely to manifest itself in the future will differ from that of the past.

To gain a perspective on the energy security issues for the remainder of this decade, we will examine how the economic and political situations have changed in recent years. Starting with an overview of the changing structure of the international oil market, we offer a broad assessment of market performance over the next several years and the implications for various energy policy strategies. We then analyze the political dimensions of the international oil trade and the prospects for oil supply disruptions. Last, we identify and assess diplomatic and military priorities. Our objective is not to specify the precise mix of policies to enhance energy security but to clarify how policy must adapt to changing realities. Although some useful lessons may be gained from the experience of the last decade, our message cautions against the presumption that history will repeat itself.

The Changing Structure of the International Oil Trade

The world oil market has entered a new phase since the traumatic events of the 1970s, demonstrating that markets adjust, however slowly and imperfectly, as participants respond to pressures and incentives. The changes do not mean that supply insecurity and occasional price shocks are worries of the past, although market instability is less likely to result from conscious economic decisions by the oil-exporting countries. The view commonly held in the 1970s, that supply shocks and higher oil prices always benefit oil-exporting nations at the expense of oil-importing nations, is no longer widely believed. Now both groups of countries stand to lose from market instability.

The broad pattern of market behavior experienced during the 1970s is unlikely to be repeated over the next several years.[1] During the 1970s interrupted oil production led to a period of hoarding and inventory building; panic buying contributed to, if it was not largely responsible for, the two major price shocks that occurred. In the case of each shock, new price plateaus were established by the actions of OPEC countries. Production

1. Although our focus is on the remainder of this decade, this and subsequent "forecasts" are intentionally stated in general terms to avoid the impression that precision is either possible or credible.

2

schedules tended to validate each new price plateau, so that the price of crude oil followed a ratchet-like path, with periods of relative stability followed by discrete jumps to new levels.

Supply shocks may still occur in the future and can be expected to cause price shocks. But these will be of smaller amplitude and shorter duration than might be expected from past experience. Both OPEC and private firms will behave differently than they did in the 1970s in producing this new pattern.

The OPEC countries as a whole are likely to find it contrary to their self-interest to validate temporary price increases, and they may even actively seek to restore normal operations in periods of instability. Future supply shocks are thus more likely to lead to an oscillation of prices than to produce the ratchet-like price path of the last decade.

These expected changes in market behavior are the result of fundamental changes in conditions that have occurred in recent years. Adjustments in world energy supply and demand and in the way international oil transactions are conducted impose new constraints and offer new incentives to participants in the market. The adjustments in the market are themselves the result of rational responses to a previous set of constraints and incentives. This evolutionary process will undoubtedly continue, making forecasts based on simplistic extrapolations of past relationships and past performance unreliable. By the same token, this evolutionary process requires a continual reappraisal of energy policy and international relations. Policies grounded on conditions extant in the 1970s are likely to be inappropriate for the 1980s and beyond.

The Evolution of Energy Supply

Evidence of the changing structure of petroleum supply is nowhere more apparent than in the OPEC countries' declining share of production (table 1). The steadily rising share of non-OPEC production is the direct result of the trend to higher oil prices, and the higher prices are in turn maintained by the willingness of OPEC to restrain output. OPEC's self-imposed restraint thus becomes an incentive for others to invest in exploration and production in new areas.

The trend away from OPEC supply may be expected to continue as long as OPEC is willing to relinquish its market share and as long as production capacity in non-OPEC countries continues to grow. Most of the new oil provinces in non-OPEC countries have higher development costs and smaller volumes of recoverable reserves and hence are brought

3

Table 1. *Crude Oil Production, 1973–83*
Million barrels per day unless otherwise specified

Year	Arab members of OPEC	Total OPEC	Noncommunist countries outside OPEC	USSR	World total	OPEC share of world production (percent)	USSR share of world production (percent)
1973	18.0	31.0	15.2	8.5	55.7	56	15
1974	17.7	30.7	14.8	9.0	55.9	55	16
1975	16.0	27.2	14.6	9.6	52.9	51	18
1976	18.6	30.7	14.8	10.1	57.3	54	18
1977	19.2	31.3	15.8	10.7	59.7	52	18
1978	18.5	29.8	16.9	11.2	60.1	50	19
1979	21.1	30.9	18.0	11.5	62.5	49	18
1980	19.1	26.9	18.7	11.8	59.5	45	20
1981	15.8	22.6	19.4	11.9	55.9	40	21
1982	11.7	18.8	20.4	12.0	53.2	35	23
1983	10.3	17.6	21.1	12.0	52.9	33	23

Source: U.S. Department of Energy, Energy Information Administration, *Monthly Energy Review* (March 1984), pp. 108–09.

into production at a competitive disadvantage relative to OPEC countries. Thus it would be easy enough for the OPEC producers to recoup their market share through a change in production strategy that relinquishes control of prices. Indeed, economic pressure is building to force a change in OPEC production strategy or, in the event that cohesion is not preserved, to force countries to act independently.

One source of pressure is the fact that reserves continue to grow in non-OPEC areas fast enough to support ever higher rates of output. Individually, the discoveries are often small and unimpressive, but in the aggregate they are changing the structure of world oil supply. In 1983, for example, fully two-thirds of all non-OPEC producing countries registered gains in output, including production from at least ten new fields in the North Sea, from individual new fields in Thailand, Brazil, and Oman, and from expansions in reserves in Angola, Cameroon, and the Congo.[2] Recent reports also offer bright prospects for important new discoveries off the north coast of Australia, off the shore of Alaska (notwithstanding the Mukluk disappointment), and in new areas of the North Sea.

For the future, unexplored regions in a variety of less developed countries offer considerable potential. These areas have gone untapped in the past for several reasons, including the reluctance of private investors to bear the financial risk and the unwillingness of host governments to design more attractive contractual relationships.[3] That situation is now changing and represents another aspect of the petroleum market in transition. As the inventory of unexplored prospects is progressively depleted, farsighted private investors are forced to pay greater attention to the less developed countries and show a willingness to bear the risk of investing in these countries. At the same time, host governments in less developed countries have learned that they must actively compete for prospective investors. As the parties move closer to workable contractual relationships, both their interests and those of the world's oil consumers will be served.

The emphasis on the growing volume of output coming into the market from non-OPEC areas should not overshadow the importance of the diverse set of interests represented by the growing number of producing countries. Within this group is a wide range of motives for developing

---

2. *Petroleum Intelligence Weekly*, vol. 22 (December 12, 1983), p. 11.
3. Harry G. Broadman, "Determinants of Oil Exploration and Development in Non-OPEC Developing Countries," Discussion Paper D-114 (Washington, D.C.: Resources for the Future, October 1983); and Raymond F. Mikesell, *Petroleum Company Operations and Agreements in the Developing Countries* (Washington, D.C.: Resources for the Future, 1984).

their resources and a correspondingly wide range of responses to changes in economic and noneconomic conditions. This diversity will contribute to overall stability in world supply by smoothing over discrete changes in individual activities. In addition, the non-OPEC share of world supply becomes more competitive as individual producers become increasingly atomistic within the group and act more like price takers. Last, the possibility of cartel action by non-OPEC suppliers fades with the number and variety of interests involved.

Other sectors of the petroleum industry are also in transition, and the move by OPEC countries into downstream activities is particularly important. Unlike their share in crude-oil production, OPEC's involvement in petroleum refining and marketing is rapidly increasing. The volume of refining capacity located in OPEC countries is expected to increase from 6 million barrels per day in 1980 to nearly 9 million barrels in 1985 and perhaps to 12 million barrels by 1990.[4] Most of the increase, moreover, will take place in the politically sensitive countries in the Persian Gulf region.

The importance of OPEC's expansion in refining capacity is heightened by the corresponding contraction under way in traditional refining centers, which cannot compete on the same terms with the oil-exporting countries. Existing capacity will therefore be displaced as new capacity comes on stream. Indeed, the process is already under way as investment in the maintenance, upgrading, and restructuring of existing capacity lags in anticipation of future developments.

OPEC's move into downstream operations, however, is a double-edged sword. On the one hand, OPEC can expect to derive substantial political and economic advantages from the move. The political advantages follow from the enhanced potential for controlling the distribution of oil among world markets. The economic advantages occur because the investment is at home, it permits use of gas that otherwise would have been flared, and it offers the potential for package deals involving crude oil, petroleum products, petrochemicals, and transportation.

The disadvantage to OPEC is that the pricing of refined products adds a new headache to the already difficult process of controlling crude-oil price differentials. The market for refined products is extremely complex and

4. Fereidun Fesharaki, "Wide Impact Seen for OPEC's Refining Push," *Petroleum Intelligence Weekly*, vol. 20, special supplement (June 22, 1981), supplement p. 1; and Hossein Razavi and Fereidun Fesharaki, "OPEC's Push into Refining: Dilemma of Interactions Between Crude and Product Markets," *Energy Policy*, vol. 12 (June 1984), pp. 125–34.

Table 2. *World Production of Natural Gas and Coal, Selected Years, 1973–82*

Product and region	1973	1975	1977	1979	1981	1982[a]
Natural gas (trillion cubic feet)						
North America	24.68	22.20	22.29	23.13	22.90	21.32
Western Europe	5.15	6.07	6.01	6.44	6.92	6.39
Eastern Europe and USSR	9.96	11.95	14.26	16.32	18.49	19.59
Middle East	0.98	1.19	1.41	1.58	1.36	1.34
Rest of world	2.38	2.71	3.34	4.40	4.86	4.96
Total	43.15	44.12	47.31	51.87	54.53	53.60
Coal (million short tons)						
North America	626	688	736	822	876	893
Western Europe	508	509	506	529	562	574
Eastern Europe and USSR	1,431	1,494	1,546	1,571	1,529	1,581
Rest of world	882	982	1,052	1,199	1,261	1,334
Total	3,447	3,673	3,840	4,121	4,228	4,382

Source: U.S. Department of Energy, Energy Information Administration, *1982 International Energy Annual* (Government Printing Office, 1983), pp. 18, 20.

a. Preliminary.

more difficult to police than the crude-oil market and hence affords a wider range of opportunities for deviating from agreed-upon prices. The incentive to cheat increases with the added investment in refining and distribution facilities and with the corresponding domestic employment involved in these operations. Individual countries faced with a decision to reduce output will be inclined to shave prices and maintain production. Since this problem will not occur in a strong and expanding market, OPEC's declining share of world oil supply takes on additional importance and adds to the pressure on OPEC to modify its market strategy.

Another source of pressure on OPEC's market strategy comes from competition with coal and natural gas. As one might expect, the rising price of oil has stimulated increased production of coal and natural gas worldwide (table 2), providing a sharp contrast to the declining trend in oil production.

Coal production and consumption, while growing, are hindered by environmental concerns and the capital costs of addressing these concerns. Natural gas represents perhaps the strongest source of competition with oil in the coming years, and the growing gas production capacity of Europe and the USSR has become especially prominent. Western European countries have yet to rely on natural gas as intensively as the United States has, but the gap will narrow in the future. European gas production, while steadily increasing over the last decade, has yet to show the effects of recent large gas discoveries in the North Sea. In addition, the Soviets are

7

poised to begin major new shipments to Western Europe through the recently completed gas export pipeline.

Soviet gas represents a major source of competition with imported oil. The Soviets have been rapidly developing their huge gas fields, which account for about 40 percent of proven and probable world reserves, and are anxious to expand export markets. The Europeans, on the other hand, started backing away from long-term commitments as the effects of lagging energy demand and competition with oil became apparent even before the gas started to flow. France, West Germany, Austria, and Switzerland originally contracted to purchase a combined total of 610 billion cubic feet of Soviet gas annually for twenty-five years, and all four subsequently decided to renegotiate their contracts. In early 1984 Italy temporarily postponed signing a contract for 240 billion cubic feet annually, while five other European countries backed away indefinitely, citing revised forecasts of energy requirements.[5]

The large amounts of new gas available to Western European consumers at current prices impose a real constraint on future OPEC pricing strategy. The potential for a major shift away from oil will make it difficult for OPEC to rationalize a higher price plateau in the next several years, for a higher oil price is precisely what is required to bring the Europeans to terms with the Soviets on arrangements for gas.

The Evolution of Energy Demand

The overall pattern of world oil demand in recent years matches that of the aggregate oil production figures cited in table 1, with consumption fluctuating during the 1973–78 period and declining steadily thereafter. The industrialized oil-importing countries' share of world oil consumption warrants particular attention. The twenty-one members of the International Energy Agency constitute most of the world's noncommunist industrial countries, and they account for 70 to 75 percent of total noncommunist oil consumption and most of internationally traded oil.[6] As a

5. Early reports indicate that Italy has been successful in negotiating better terms than the four countries under contract. See *Petroleum Intelligence Weekly*, vol. 23 (June 18, 1984), pp. 2–3.

6. The IEA was formed by twenty-one of the twenty-four members of the Organization for Economic Cooperation and Development (not participating were France, Finland, and Iceland) to provide a coordinated defense against oil supply disruptions. For details on the formation, structure, and objectives of the IEA, see Mason Willrich and Melvin A. Conant, "The International Energy Agency: An Interpretation and Assessment," *American Journal of International Law*, vol. 71 (April 1977), pp. 199–223.

Table 3. *Characteristics of Energy Consumption in IEA Countries,
1973, 1980, 1983*[a]

Millions of tons of oil equivalent unless otherwise specified

Characteristic	1973	1980	1983[b]
Total primary energy consumption (TPE)	3,324.4	3,562.7	3,359.0
Oil consumption	1,709.6	1,666.6	1,460.6
As a percent of TPE	51.4	46.8	43.5
Net oil imports	1,168.1	1,045.5	734.9
As a percent of TPE	35.1	29.4	21.9
As a percent of oil consumption	68.3	62.7	50.3

Source: Organization for Economic Cooperation and Development, International Energy Agency, *Energy Policies and Programmes of IEA Countries, 1983 Review* (Paris: OECD, 1984), p. 13.

a. The IEA includes twenty-one of the twenty-four members of the OECD; excluded are France, Finland, and Iceland. Over the 1973–83 period, the IEA countries accounted for 70 to 75 percent of the total world consumption of oil.

b. Preliminary.

group, they are highly dependent on oil imports. Changes in the level and composition of energy consumption in these countries will largely determine worldwide trends in oil prices and the overall political importance of oil trade.

Since 1973 there have been some notable changes in the energy consumption patterns of countries in the International Energy Agency (table 3). Consumption of total primary energy (composed largely of oil, gas, and coal) continued to rise between 1973 and 1980, though at a slower pace than in previous decades, and declined dramatically after 1980 as a combined result of higher prices and slow economic growth. Over the entire period, the share of oil in total energy consumption declined, slowly at first and sharply after 1980. By 1983 total energy consumption was not much higher than in 1973, while oil consumption was 15 percent below the 1973 level.

The seemingly modest decline in the share of oil in total energy consumption between 1973 and 1980 represents a reversal in the trend established in previous decades and is all the more remarkable in that it occurred in a period of continued economic growth. Large-scale substitution of other fuels for oil takes many years to accomplish because of the costs of altering the capital stock (which determines energy requirements) and because increased production of coal and natural gas takes time.

Progress in the transition away from oil is also reflected in the International Energy Agency countries' declining dependence on oil imports. From 1973 to 1983 their net oil imports declined from 35.1 to 21.9 percent of total energy consumption and from 68.3 to 50.3 percent of total oil consumption. Although these figures indicate that the noncommunist

9

industrial economies continue to depend heavily on oil as a source of energy and on imports as a source of supply, the direction of change exerts an important influence on the market. For oil-importing countries, the trend away from oil means that the potential economic dislocations of an oil-price shock have eased along with the degree of panic likely to be associated with a sudden oil-supply shock. For oil-exporting countries, the downward trend creates slack in existing oil-production capacity and increases competition among those seeking to maintain export revenues. While the incremental reductions are small in relation to the volume of consumption and trade, their importance is magnified in determining the competitive atmosphere in the market.

Naturally, the decline in oil consumption will slow if the price of oil remains stable and will reverse direction if the price of oil falls for an extended period. There are, however, powerful forces that will work to sustain the momentum toward conservation over the next several years. One source of inertia is the slow process by which capital stock is replaced. Almost all energy is consumed in association with a unit of capital equipment, and the rate of consumption is determined by a combination of the fuel efficiency of the capital stock and its rate of utilization. An increase in the relative price of oil will alter investment decisions in favor of oil-efficient capital goods, but with a considerable time lag because of the slow turnover in equipment. Consequently, the changes initiated during the 1970s will continue to come on stream over the next several years.

The income effect on energy demand manifests itself through the rate of utilization of the existing capital stock and through the demand for new capital formation. These two influences generally exert opposite, though not equal, pressure on energy demand. Slow economic growth reduces current energy demand by slowing the rate of utilization of existing capital, but it also slows the transition to more energy-efficient capital stocks by reducing the need for new capital formation. The opposite occurs with higher rates of economic growth.

These separate influences on demand may be illustrated in the case of gasoline, where gasoline demand is a product of the fuel efficiency of the automobile fleet, the size of the fleet, and the average distance traveled per car. The average efficiency of new cars entering the automobile fleet has improved steadily since 1973, and this trend may be expected to continue (table 4). The overall efficiency of the automobile stock will continue to improve as older cars are replaced even if improvements in new car efficiency slow to a halt. Average distance traveled per car has also

Table 4. *Characteristics of Automobile Energy Demand in Selected Countries, Selected Years, 1973–85*

Characteristic	1973	1978	1980	1981	1985[a]
New car fuel efficiency (liters per 100 km)					
West Germany	10.3	9.6	9.0	8.7	8.1–8.6
Japan	10.5	n.a.	8.3	8.1	7.8
United Kingdom	11.0	10.1	9.6	9.6	9.1
United States	18.8	13.1	12.0	10.8	8.6
Average distance traveled per car (1,000 km)					
West Germany	13.5	13.2	12.8	11.9	. . .
Japan	15.2	10.4	10.2	10.0	. . .
United Kingdom	14.2	15.4	15.2	15.1	. . .
United States	16.1	16.1	14.7	14.5	. . .

Source: OECD, IEA, *Energy Policies and Programmes of IEA Countries, 1982 Review* (Paris: OECD, 1983), p. 30.
n.a. Not available.
a. Expected.

declined in recent years as a result of higher fuel prices and sluggish growth of income. Distance traveled per car is a more volatile variable than the capital stock and may be expected to turn around on short notice with a resurgence in economic growth and stable gasoline prices.

Transportation demand accounts for slightly more than half of total oil consumption in the noncommunist industrial countries, and there are few practical alternatives to petroleum to satisfy transportation needs. Consequently, transportation demand for oil is less flexible than many other end uses, and conservation is more narrowly achieved through changes in fuel efficiency and utilization rates. Thus transportation demand for fuels is especially important in assessing energy balances. As the figures show, however, even the transportation component of oil demand is highly responsive to upward movements in oil prices, and past price increases will act as a drag on demand for years to come.

Industrial demand for petroleum products has exhibited greater flexibility in response to changes in relative energy prices. For example, table 5 shows the remarkable decline in residual fuel-oil consumption in six industrial countries. Residual fuel oil is a boiler fuel in large industrial and electricity generating facilities, and the decline reflects both the elimination of oil-fired capacity and the alteration of equipment to permit fuel switching.

Another powerful support for petroleum conservation, even with stable or temporarily falling oil prices, is the uncertainty associated with oil supplies and prices. Investment in energy-using capital stocks is generally costly and long term, with decisions based on expectations about the

11

Table 5. *Residual Fuel Oil Consumption in Selected Countries, 1979–83*
Thousand barrels per day

Country	1979	1980	1981	1982	1983[a]
United States	2,826	2,508	2,088	1,668	1,377
Japan	1,556	1,345	1,159	1,015	938
Italy	727	723	669	602	560
France	531	455	336	277	194
United Kingdom	500	348	284	294	223
West Germany	409	369	290	258	208
Total	6,549	5,748	4,826	4,114	3,500

Source: *Petroleum Intelligence Weekly*, vol. 23 (February 13, 1984), p. 4.
a. First nine months.

future in addition to contemporary market conditions. Because motives and expectations are shaped in part by past experience, instability in the oil market in the 1970s continues to provide the incentive for investments in the 1980s that favor oil conservation and substitution capability. Thus a lengthy period of price stability is required to shake the legacy of the past.

Changes in Market Institutions

The world will not always have the comfort of the substantial excess oil-production capacity that it has had in the early 1980s. Demand can be expected to grow with rising incomes even though energy is used more efficiently. If demand fails to absorb excess capacity fast enough, producers will slow investments in exploration and development of new reserves and the volume of excess capacity will fade. Whether the impetus comes from the pressure of demand or of production costs, sometime in the near future (perhaps the early 1990s) there will be a tightening of the market.

How will the international oil market perform under tighter conditions? Short-run price stability is critically dependent upon the demand for inventories of crude oil and refined products; the demand for inventories is the most volatile component of oil demand (as opposed to oil consumption). Stock building in a crisis can exacerbate the price shock, as indicated by experience after the Iranian revolution, while a drawdown can moderate prices, as indicated by the response to hostilities between Iran and Iraq. For this reason, the factors that influence inventory demand are important to market stability.

Private inventories are held for several reasons, including the need for working balances as well as the desire for precautionary and speculative stocks. The volume of working stocks is a fairly stable component that varies in relation to the volume of consumption, including seasonal fluctuations and long-term trends. The demand for precautionary and speculative inventories is more critical because it influences the market during periods of supply instability. Risk aversion and speculation are often accorded the blame for the surge in oil prices that followed the Iranian revolution.[7] For reasons discussed below, inventories are likely to respond quite differently to another supply shock.

Until very recently much of the world's oil reached consumers through a highly integrated distribution chain, with each stage of the process from extraction to refining and marketing linked to specific sources of supply and to specific market outlets. Relatively few open-market transactions occurred along this chain, and entry into the market was difficult because of high capital costs and long-term contracts. Even the advent of nationalization and the rise of OPEC market power after 1970 did not immediately alter traditional ways of doing business: long-term contracts merely replaced concessions as the means of securing access to crude oil, while downstream distribution systems worked as before. Except for competition in selling products to consumers and in bidding for long-term crude oil contracts, the market was characterized by a small number of exchanges of crude oil and products among firms.

Long-term contractual arrangements provided one means of establishing order in the market and of enhancing security of supply to individual firms. In this environment, the spot market performed the minor function of assisting refiners in balancing their input and output mixes; firms relied on their own distribution channels for all but these marginal transactions. The specialized role accorded to spot transactions thus effectively separated the spot market from the rest of oil trade. The system also required sizable inventories of feedstocks and refined products to smooth over erratic changes in supply and demand and to provide a cushion in the event of a disturbance.

After the turmoil created by the Iranian revolution in 1978–79, the oil market began to change rapidly. The distribution process began to rely less on long-term contracts and more on spot transactions. The transition was

---

7. See Philip K. Verleger, Jr., *Oil Markets in Turmoil: An Economic Analysis* (Ballinger, 1982); Albert L. Danielson and Edward B. Selby, "World Oil Price Increases: Sources and Solutions," *Energy Journal*, vol. 1 (October 1980), pp. 59–74; and Douglas R. Bohi, "What Causes Oil Price Shocks?" Discussion Paper D-82S (Washington, D.C.: Resources for the Future, January 1983).

also marked by changes in decisions concerning the role of inventories, the volume and mix of refining capacity, and reliance on marketing outlets. These adaptations signaled fundamental changes in market behavior.

Since 1980 the market has come to rely less on long-term contracts, because these instruments have proved to be unreliable at precisely the time they are most needed. Contract terms have been subject to immediate revision—sometimes even retroactively—when market conditions change, thus failing to provide the security for which they were intended. The contracts served more as an expression of intent and were adjusted as the interests of the parties changed with market conditions. While the advantages of long-term contracts proved illusory to buyers, definite disadvantages began to emerge during the critical 1980–81 period. Posted prices for crude oil failed to adjust to developing excess supply conditions and were well above the value of crude oil (netbacks) dictated by market-determined prices of products. Consequently, firms attempting to maintain their integrated channels of supply could not compete with prices of feedstocks and products moving through the spot market. For the first time, the spot market began to emerge as a full-fledged market with an increasing volume of transactions and a growing number of traders competing in all phases of the business.

All segments of the oil industry have been motivated to adjust to the competition introduced by growing spot-market activity. The various stages of integrated operations were forced to compete with prices established by open-market transactions. The squeeze on refinery profits and competition for customers, for example, forced integrated firms to pay closer attention to margins involved in making their own products versus buying on the spot market. Competition among alternative sources, in turn, tended to break down the traditional links among consumers, distributors, refiners, and producers. Even private investors outside the industry gained access through spot-market transactions, buying and selling spot contracts for arbitrage profits or investing in inventories for speculative gains.

The growth in spot-market activity also affected the motives for private inventory demand. The need to hold stocks has been reduced because the spot market enhanced access to alternative sources of supply; the willingness to hold stocks has declined because the competitive pressure exerted by spot transactions made it difficult to pass through to customers the carrying costs of inventories. For these reasons, average stock levels per firm may be expected to decline. Total private inventories need not fall, however, because easier access to the market through spot and futures transactions encourages an expansion in the number of firms holding petroleum stocks.

The emergence of the spot market has also affected the government-controlled oil companies because, on the buying end, governments of oil-importing countries have come to learn that bilateral relationships with oil-exporting countries are not necessarily advantageous.[8] The move in the 1970s toward bilateral trade agreements has been reversed, and several countries have begun to rely heavily on the spot market as the primary source of supply. On the selling end, oil-exporting countries have reduced their dependence on contract deals by creating marketing subsidiaries to sell through the spot market. Even the OPEC countries, which have traditionally ignored the spot market and in the past have even threatened sanctions against firms reselling contract oil on the spot market, have been forced to enter it. This became necessary as buyers shifted from long-term contracts and as competing producers responded to the shift. As Iran, Iraq, and Libya moved aggressively into the spot market, it was inevitable that Saudi Arabia and others would soon follow.

The shift away from official bilateral trading relationships, combined with the active entrance of some OPEC members into the spot market, will doubtless help reduce the political element involved in oil transactions. At the same time, the market has become less segmented according to geographic area, making it easier for the market to allocate supplies according to willingness to pay. Regional price differentials need not get far out of line to induce a redistribution of supplies, so the market can more effectively smooth out regional disturbances in supply and demand.

Another major corollary of the transition to a more competitive structure is the recent emergence of a futures market in crude oil and refined products. Organized futures markets cannot exist when prices are unable to fluctuate with underlying economic conditions and when transactions are difficult to arrange. That futures markets became viable for the first time in the early 1980s is an indication of how much the oil market has changed.

The influence of futures markets in the evolution of the oil market remains to be seen, but the potential is substantial. The essential role of the futures market is to transfer price risk to parties willing to bear the burden and in the process to improve the efficiency of the oil market and the fungibility of oil trade. The role of an organized futures market in this

8. Concerns about the proliferation of national oil companies before 1980 are given in Brian Levy, "World Oil Marketing in Transition," *International Organization*, vol. 36 (Winter 1982), pp. 113–33.

respect is analogous to the role of money in facilitating trade.[9] In the absence of a futures market, hedging against price risk is achieved through a forward contract that specifies the terms of a future delivery at a price determined today. The transaction is similar to a barter arrangement because the parties must have corresponding interests and must know and trust each other and, importantly, because the forward contract itself is not easily exchangeable with another party. In contrast, all futures contracts traded on an organized market are perfect substitutes, and their validity is independent of the identity of the parties to the contract. This is because the contract is a creation of the clearing house of the market, and each buyer of a contract obtains a liability of the clearing house that is matched by an offsetting asset represented by the sale of the contract. Consequently, the contract is highly negotiable and, like currency, facilitates trade by lowering transaction costs.

The introduction of organized futures markets for oil and refined products should enhance the efficiency of transactions beyond that achieved by an active spot market alone. Part of the efficiency gain derives from the additional information about market conditions conveyed through spot and futures prices, the added speed with which price differentials induce reallocations of supplies, and the reduction of costs. An example of these gains is the substitution of futures contracts for oil inventories that are held for the purpose of hedging against unfavorable price movements or for the purpose of speculating on the future. Since the exchange of pieces of paper is cheaper and easier than handling physical stocks of oil, the futures market increases overall efficiency. By the same token, since the transaction cost of adjusting a futures position is less than that of altering the volume of physical stocks, inventory demand should become less sensitive to fluctuations in the price of oil. This follows in part because inventory demand will be determined more by the convenience benefits of working stocks and less by speculative motives. In addition, those who can least afford to be caught short in a crisis, along with those most averse to risk, are more likely to cover themselves with contract purchases. Consequently, panic buying in a crisis will be mitigated to some extent by the added insurance purchased through the futures market.[10]

9. See Lester G. Telser, "Why There Are Organized Futures Markets," *Journal of Law and Economics*, vol. 24 (April 1981), pp. 1–22.

10. Aggressive bidding will not be eliminated because, among other things, the marginal convenience benefits of inventories rise as the stock level falls. Refiners, for example, will be willing to increase their bids for additional stocks as their shutdown point is approached. Similar incentives will motivate other consumers.

The moderating effects of spot and futures markets on panic buying are limited, to be sure, because in severe and extended crises trading will be sharply curtailed if not suspended. The value of spot and futures markets is in moderating the demand shock that can accompany relatively mild supply disruptions: it forestalls price bubbles created by perverse expectations and quickly redistributes access to supplies according to willingness to pay. In view of the experiences of the 1970s these adjustments in the way the market will respond to minor supply crises should not be underestimated.

Implications for Market Stability and Energy Policy

In the preceding discussion we have described a number of factors that will influence the long-term trend of future oil prices and the short-term behavior of oil prices in the event of a supply disruption. While it is not possible to forecast the precise path of prices, something can be said about the prospects for market stability and the reasons why prices are likely to behave differently in the future than they did in the past. These implications are derived from changes in incentives that shape OPEC and private market behavior. These changes also require that aspects of energy policy based on earlier market behavior be reexamined.

Future Price Behavior

In considering future prices it is useful to examine the behavior of OPEC separately from that of private agents. OPEC may be viewed as capable of production decisions that will initiate disturbances (insofar as disturbances are the result of conscious strategy) and that will establish the price floor, while private agents respond to market conditions in ways that reinforce or counteract OPEC production and pricing decisions.

OPEC strategy is constrained by the increasing trend of production outside OPEC, the declining trend in oil consumption, and the fact that both trends are the result of past OPEC decisions. As shown in table 6, total OPEC revenues have dropped sharply since 1981, following OPEC's decision to limit production and defend the price increases established in 1979–80. The decline in total revenues following the price increases is convincing evidence of a fairly elastic demand for OPEC oil. With OPEC's market share reduced and increasingly sensitive to changes in

17

Table 6. *Changes in OPEC Oil Revenues, 1981–83*

Billions of dollars

Country	1981	1982	1983[a]
Saudi Arabia	113.2	76.0	47.4
United Arab Emirates	18.7	16.0	11.3
Kuwait	14.9	9.5	10.3
Iraq	10.4	9.5	8.3
Iran	9.3	16.4	19.6
Nigeria	16.7	13.1	10.4
Libya	15.6	14.0	11.1
Venezuela	17.4	13.5	13.4
Indonesia	14.4	12.7	10.0
Ecuador, Gabon, and Qatar	7.9	5.8	5.8
Total	249.2	195.0	154.0

Source: *Petroleum Intelligence Weekly*, vol. 23 (April 23, 1984), p. 5.

a. Estimates based on OPEC production of 17.5 million barrels per day.

consumption and non-OPEC supply, further reductions in OPEC oil revenues may be expected to accompany higher price levels.

The implications for OPEC's behavior are that it will not be in the oil cartel's economic interest to initiate or sustain an increase in oil prices, and individual countries will be disinclined to follow a cooperative strategy that sacrifices their market share. Unlike the situation in the 1970s, when the interests of exporting and importing countries seemed to be diametrically opposed, it now appears that OPEC and the importing countries will both lose if prices rise. Similarly, in the event that a supply disturbance occurs for noneconomic reasons, it is not in OPEC's interest to capitalize on the price shock and try to establish a new price floor as in the 1970s.

The increased volatility of OPEC earnings in response to changes in the price level suggests that OPEC has a tangible economic incentive to help stabilize the market. Even temporary fluctuations in the price may be contrary to its interests. Apart from their damaging effect on long-run demand for OPEC oil, price fluctuations increase the difficulty of maintaining cooperation in production and pricing agreements, particularly in the downward phase of the price cycle. In addition, price fluctuations and their attendant revenue fluctuations make internal budget planning difficult, cause turmoil in planning domestic development projects, and create political tension. Consequently, it is reasonable to imagine that most OPEC members will recognize their self-interest in stable prices in coming years.

Whatever the economic incentives that underlie OPEC's motives and actions, supply disruptions and their resulting price shocks can be expected to occur, as will be emphasized below in the discussion of political tensions. The severity of a disruption, however, will depend not just on OPEC's behavior but to a significant degree on private behavior, which can mitigate or exacerbate price shocks. Recent developments in the oil market suggest that private responses will have less perverse effects than those experienced during the 1970s.

The market is less susceptible to the trauma of a disruption than in the past decade. The adjustments in transactional arrangements described above mean that commodity flows are less likely to become frozen within traditional channels and that the price will more efficiently seek a market-clearing level. The spot market is less likely to dry up as it has in past periods of supply uncertainty, and arbitraging between contract prices and spot prices is more likely to remain active. As a consequence, large disparities between spot and contract prices are unlikely to occur, eliminating the contribution of such differentials to panic buying behavior.

The market should also work faster and more efficiently in allocating supplies among firms and regions because fewer rigidities need to be overcome in bidding for available supplies. Firms and regions relatively more dependent on the spot market will be less severely affected in a tight market, and those with long-term contracts affected by the crisis can expect to turn to the spot market. An active spot market will thus spread the shortage across buyers more quickly and more evenly and, as a result, reduce panic bidding motivated by the need to overcome market rigidities.

Inventory demand will also be less volatile in periods of supply uncertainty. This may be expected in part because of the emergence of spot and futures markets and their dampening effect on hedging and speculative behavior. Also important is the fact that private agents have gained valuable experience from past periods of instability. In particular, firms learned in 1980 that capital losses on inventories are possible and that there is a point at which the risk of overbuilding inventories exceeds the risk of being caught short.

If this lesson has not been widely accepted yet, it will be reinforced as prices demonstrate a tendency toward cyclical behavior rather than follow the ratchet-like path of the 1970s. As this expectation about price behavior becomes more widely adopted, firms will perceive that they can ride out the difficult phase of a crisis and can be less risk-averse in their inventory strategies. In sum, firms will tend to add less to stocks in the early phase of a crisis and will become more inclined to sell from inventories as the price rises, thus moderating the price peak and speeding the decline.

19

Implications for Energy Policy

The international oil market has adapted to the traumas of the 1970s with a more flexible system of conducting transactions, more diversity in the distribution of oil production, and more efficiency in oil consumption. In short, the market is more resilient now and oil prices may be expected to behave differently as a consequence.

A look at the last decade shows that energy markets do work, if at times slowly and imperfectly. The widespread pessimism prevalent in the 1970s, that energy supply would not increase and energy demand would not fall at higher prices, has proved to be unfounded. In this connection, it is well to remember that many energy policy initiatives of the last decade were supported by gloomy forecasts of future energy supply and demand imbalances.[11] In the most naive forecasts, the market was characterized as incapable of adjusting to increasing energy scarcity or of providing sufficient advance warning of impending shortages to enable society to make the necessary adjustments. The presumed link between the volume of energy consumption and the growth of economic welfare, on the one hand, and the presumed physical limits of available energy resources, on the other hand, created a sense of urgency to implement policies that would speed the transition from conventional to new energy resources. The experience of the last decade should help put these arguments to rest.

For some, skepticism about market solutions lies not so much in doubts about their technical efficiency as in the political repercussions of allowing markets to work.[12] For one thing, markets take time to adjust, and the time lag can be a politician's nightmare. It is difficult for governments to resist public pressure to do something, even if intervention may be counterproductive in the long term. In addition, market adjustments inevitably arouse arguments about equity. This concern is usually expressed in public debates about the burden placed on the poor, about consumers versus producers, and about regional shifts in income. These issues necessitate government involvement in energy markets, and allowances must be made to sacrifice efficiency on behalf of equity.[13]

11. The most prominent among these forecasts is the Club of Rome Report published in Donella H. Meadows and others, *The Limits to Growth* (New York: Universe Books, 1972).
12. George P. Shultz and Kenneth W. Dam, *Economic Policy Beyond the Headlines* (Norton, 1977), pp. 179–97.
13. The issues are complex and the solutions are not obvious, as indicated by Hans H. Landsberg and Joseph M. Dukert, *High Energy Costs: Uneven, Unfair, Unavoidable?* (Johns Hopkins University Press for Resources for the Future, 1981).

There are, in addition, legitimate efficiency arguments for government intervention in energy markets. The most forceful of these arguments hold that private initiatives are inadequate in inducing optimal adjustments to energy security problems; these arguments call for government support of energy research and development, for government-controlled stocks of oil for use in an emergency, for international coordination of energy policies, and for an oil import tariff. The first three of these programs are under way in the United States, while an import tariff has been widely supported as an energy security measure.

The following sections will examine each of these policies in light of the structural changes in energy markets described above. Typically, the urgency of an energy policy decision and the appropriate choice among policy options depend upon assumptions about the way markets are expected to perform. Because of the changing structure of energy markets, these decisions are subject to continual reinterpretation.

Energy research and development. The basic argument for government support of energy R&D is valid: information is created that is a public good in the sense that it benefits all of society without adequate compensation to private investment. As a consequence, private investment in new technology that permits the transition from conventional to new energy sources will be below the social optimum, and additional government support is required to make up the difference.

The adjustments in energy markets during the last decade provide a perspective on the appropriate focus of R&D efforts. While no one can be certain about how the energy requirements of distant generations will be satisfied, the near-term energy situation is less critical than once thought, and the need for government intervention is less urgent. Between the problem of short-term adjustments to oil-supply disruptions and the very long term problem of a transition to new technologies, the market has demonstrated the capability of responding to changing needs with a minimum of government intervention. The fact that the market works does not argue for complacency about the more distant future but only that society has more time to design and implement the appropriate energy policy strategy.

This line of reasoning suggests that the government's efforts to promote energy research and development are best directed toward the very long run technological issues; that is, toward long-term research and away from near-term development. Technologies that are on the verge of becoming commercial are neither required nor useful as targets of government support: they are not required as a supplement to the adjustments the market

21

will make in any case, nor will they solve the adjustment problems that will be required further down the road.

More specifically, the activities originally envisaged for the Synthetic Fuels Corporation are less important now than they were perceived to be when that institution was created. In promoting the development of existing technologies that offer the earliest promise of commercial profitability, the Synthetic Fuels Corporation is focusing on the wrong objective.

For similar reasons, the limits on nuclear power in the United States are now less of a hindrance in meeting near-term energy requirements. Time is available to develop new technologies and institutions that will address existing shortcomings. A temporary hiatus could be beneficial if it helps dissipate vested interests in the current structure of the industry and permits fresh approaches that are necessary to make nuclear power more acceptable to society as a component of supply in the future.

Strategic petroleum reserve. There is a well-established need for government-controlled stocks of oil to ameliorate the costs of a disruption that will not be addressed through private incentives for inventory holding.[14] More controversial is the appropriate strategy for drawing on the strategic petroleum reserve during an emergency. The optimal strategy depends on expectations about the nature of the emergency and, importantly, on expected private-sector responses to the emergency.[15]

In 1984 the Reagan administration committed itself to a strategy of using the strategic reserve early in a supply emergency, reversing the previous unofficial policy of drawing on the reserve only as a last resort. The stated reason for this strategy of early use is to dampen price increases; for this to occur, the market needs to know in advance how the reserve will be used. This reasoning is sound, particularly in view of widespread concern that the government would be reluctant to use the reserve at all.

There is little to lose and potentially much to gain from a modest drawdown of the reserve early in a disruption, assuming that the action will have an important psychological effect on the market. At question is the wisdom of a quick and *massive* drawdown, for in a prolonged interruption the psychology of the market may work against stabilization. This

14. For details, see Douglas R. Bohi and W. David Montgomery, *Oil Prices, Energy Security, and Import Policy* (Washington, D.C.: Resources for the Future, 1982), chap. 6, and the references cited therein.

15. A discussion of public-private interactions is given in W. David Montgomery and Michael A. Toman, "Strategic Oil Stockpiling: The Implications of Public-Private Inventory Interactions and Macroeconomic Disruption Costs," Discussion Paper D-82L (Revised) (Washington, D.C.: Resources for the Future, April 1984).

22

will occur in much the same way as speculation against fixed exchange rates after the central bank has depleted its foreign reserves. Once the inevitable becomes obvious, the weight of the market will move against the stabilization strategy.

The argument in favor of a massive drawdown, which may be termed a "deterrence strategy," assumes that world oil prices will follow the ratchet-like path exhibited during the crises of the 1970s. If the premise is correct, the deterrence strategy promises substantial benefits that would overwhelm the risks of an early stock drawdown. Preventing a jump in the long-run price path would result in two benefits: the cumulative savings in the current and future costs of oil and the corresponding savings in adjustment costs in the economy that would be caused by sudden oil price changes. Consequently, the deterrence strategy is a forceful argument for a rapid drawdown of the reserve.

The presumed benefits are exaggerated, however, if the market behaves differently from the assumed historical pattern. As argued earlier, OPEC is in a weaker position to defend a higher price floor during the next several years, and supply disturbances are less likely to generate perverse responses by the private sector. The expected effect of a disturbance on oil prices will be a cyclical upswing followed by a return to precrisis levels once the disturbance has ended.

In this view of market behavior, a more conservative strategy for using the petroleum reserve may be warranted. Instead of a massive drawdown, it may be prudent to draw down slowly while observing how the crisis unfolds and how the market responds. Although a drawdown may be warranted, it should follow a strategy of arbitraging between normal and disrupted prices to help smooth out the price cycle rather than preventing a new price plateau. The risks of following this course are small, and the advantage of gaining additional information about the severity and duration of the crisis could be large.

International policy coordination. The potential benefits that may be derived from coordinating energy security policies among the oil-importing countries are large.[16] By cooperating in the use of oil inventories or in the application of demand restraint measures, oil-importing countries can reduce the dislocations caused by supply crises more effectively than

16. Analyses are provided in Hung-po Chao and Stephen Peck, "Coordination of OECD Oil Import Policies: A Gaming Approach," *Energy: The International Journal*, vol. 7 (February 1982), pp. 213–20; and William W. Hogan, "Oil Stockpiling. Help Thy Neighbor," *Energy Journal*, vol. 4 (July 1983), pp. 49–71.

through independent actions. Notwithstanding the potential benefits, however, it is difficult to achieve agreement on a cooperative strategy.

U.S. energy policy is coordinated with that of other industrial countries through the International Energy Agency. Because it was created in the wake of the Arab oil embargo in 1974, the agency naturally attempted to counterbalance OPEC's market power. The oil-importing countries saw their interests and those of the oil-exporting countries as mutually incompatible, and they considered cooperation among the importing countries necessary to counter the common interests of the exporting countries.[17] This premise may be outdated in view of the changes in the market conditions facing OPEC, as described earlier. The potential for cooperation among both oil-importing and oil-exporting countries exists because both groups have a common economic interest in stabilizing the price of oil.

The principal mechanism of cooperation established by the International Energy Agency to deal with a supply emergency is the oil-sharing agreement. According to the agreement, members of the agency stand ready to reduce consumption and share available oil supplies in order to distribute a shortage across member countries in proportion to historical consumption levels. The need for the sharing arrangement was based on experience and on a fundamental distrust of market allocations. The agreement would serve to counter attempts to embargo supplies to individual countries, because the market could not be relied upon to allocate the selective embargoes across all importing countries in an equitable way.

More recent interpretations of the agreement recognize the ultimate fungibility of oil across the world's markets but emphasize the potential advantages of oil sharing in dampening panic buying in countries disproportionately affected by a disruption.[18] This view focuses on rigidities in the distribution of available supplies that prevent rapid reallocations according to willingness to pay. Those forced to bid more aggressively for their share of available supplies will cause the price to spiral higher than necessary. Oil sharing would benefit all consumers, therefore, to the extent that it serves to forestall panic buying and reduces the amplitude of a price shock.

In view of the changes in the world oil market described earlier, the potential benefits from oil sharing are lower today than during the 1970s.

17. Willrich and Conant, "International Energy Agency," pp. 199–203.
18. For additional discussion, see Douglas R. Bohi and Michael A. Toman, "Oil Supply Disruptions and the Role of the International Energy Agency," Discussion Paper D-82Y (Washington, D.C.: Resources for the Future, May 1984).

24

The usefulness of the oil-sharing agreement is reduced because the market works better today than it did before, with fewer rigidities to hinder the redistribution of supplies. This is not to say that demand restraint is an ineffective strategy during an emergency but only that the oil-sharing plan stops short of reaping the full benefits that can be achieved because it does not take advantage of the power of demand to control the price of oil.

Demand restraint in the midst of a disruption is a two-edged sword, however, and is recommended only if a substantial share of consuming countries take like actions. The advantage of demand restraint is that, with highly inelastic supply and demand for oil in a disruption, a reduction in demand can yield a substantial reduction in the world price and forestall the transfer of income to oil-exporting countries. The disadvantage is that excise taxes or import tariffs necessarily raise the domestic price above the world level.[19] This additional cost is collected by home governments and can be rebated to constituents, but the prospect of raising prices at a time of world shortage is difficult to accept. This is where cooperation is important. If participation is broad, the world price can be constrained within acceptable bounds at relatively modest tax rates. Thus, with little additional burden on domestic consumers, importing countries could reap substantial savings on their import bills.

There are also benefits to be gained from coordination of stockpile drawdown strategies among members of the International Energy Agency, and effective stockpile management could serve as a substitute for demand restraint. Unfortunately, few members of the agency have discretionary stocks of oil under government control, and prospects are dim that this option will become a viable alternative to demand restraint. Demand restraint, in contrast to inventories, does not require large public investments in advance of an emergency.

An often-cited disadvantage of demand restraint is that it invites retaliation from oil-exporting countries, which could exacerbate a crisis. This argument had merit in an earlier era, but such retaliation need not occur today. If, indeed, both groups of countries have a mutual interest in price stability, perceptions about demand restraint can be cast in a different light. In fact, a combination of demand restraint by importing countries

---

19. Another instrument would be an import quota, but a quota is generally less efficient than a tax, and the method of allocating quota licenses introduces endless opportunities for discrimination among interest groups. For an analysis of U.S. experience with an oil-import quota, see Douglas R. Bohi and Milton Russell, *Limiting Oil Imports: An Economic History and Analysis* (Johns Hopkins University Press for Resources for the Future, 1978).

and a production surge by exporting countries would stabilize the market with less burden on either group.

Oil-import tariff. As noted above, the argument for imposing an import tariff in the midst of an oil-supply disruption has merit only in the context of joint actions among member countries of the International Energy Agency. If the United States were to act alone, a "disruption tariff" would impose unacceptable adjustment costs on the American economy. During normal supply periods, on the other hand, a modest import tariff on crude oil and petroleum products has been widely recommended whether the United States acts alone or in cooperation with other importing countries.[20] This section considers whether the recommendation is strengthened or weakened by recent changes in oil market conditions.

The argument for a tariff during normal supply conditions rests on the distinction between the costs of imports to private individuals versus the costs of imports to the U.S. economy in general. When costs to the economy (or social costs) exceed those recognized in private decisions, aggregate imports will be too large. A tariff on imports is therefore recommended as an effective way to equate private and social costs and to reduce the volume of imports toward the social optimum.

There are two principal reasons why social costs may be expected to exceed private costs. The first, referred to as the monopsony buying power argument, concerns the effect of aggregate U.S. import demand on the world price. The United States accounts for approximately 25 percent of worldwide imports of oil, a share of sufficient magnitude for changes in U.S. demand to have a direct impact on the world price level. Individual private importers, on the other hand, constitute a small fraction of total U.S. demand, and they cannot discern the effect of an increase in their purchases on the price level, even though, in the aggregate, their actions will drive up the world price. When the price of oil rises because of these incremental purchases, the total cost of all imports will increase. The increase in the total cost of imports (called the marginal social cost of imports) will exceed the price of the last increment to demand (the marginal private cost) because of the additional cost imposed on imports up to the last increment. The effect of demand on price would be taken into account in determining the level of imports if the nation were to act as a single purchaser (that is, a monopsonist) rather than as a collection of individual private importers acting independently.

---

20. See Bohi and Montgomery, *Oil Prices*, and the literature cited therein.

A second source of discrepancy between private and social costs of imports concerns the risks imposed on the economy by dependence on oil imports. In large part these risks are recognized and adjusted to by the private sector, and to that extent there is no discrepancy between private and social costs of imports. A discrepancy occurs because the economy cannot adjust smoothly to a rapid increase in the price of oil. Rather, the adjustments are typically accompanied by unemployment, inflation, and loss of productivity, all of which reduce potential output and lower economic welfare. The reduction in welfare is a cost that can be addressed in advance of a disruption by reducing the degree of dependence on oil, as achieved by a tariff on oil imports during normal supply conditions.

To consider the arguments for an oil-import tariff in light of recent changes in oil-market conditions, one must focus on the relationship between world oil supply and the level of demand.[21] The degree of responsiveness of supply to a change in demand is critical to the tariff's effectiveness in lowering the world price. If supply is unresponsive to a reduction in demand—that is, if production remains largely unchanged—a tariff will be more effective in lowering the world price and will achieve greater benefits to oil-importing countries by lowering their import bills. The oil-exporting countries will absorb a larger share of the tariff through lower oil prices.

In today's environment of substantial excess production capacity, a drop in demand would impose a difficult choice on OPEC countries: establish tighter production quotas with lower revenues, or maintain production and risk a reduction in the price to a new market-clearing level.[22] Compared with the sellers' market of earlier years, it would be more difficult to achieve agreement among the members of OPEC to tighten production quotas now. Hence, the prospects of a significant decline in the world price in response to an oil-import tariff are brighter today than at any time during the turbulent 1970s, when such a policy was recommended as a step toward energy security.

21. This relationship is also of critical importance in determining the optimal level of the tariff. For a review of tariff estimates under different assumptions, see Harry G. Broadman, "Review and Analysis of Oil Import Premium Estimates," Discussion Paper D-82C (Washington, D.C.: Resources for the Future, December 1981).

22. Oil production in non-OPEC areas would not be immediately affected by a decline in demand and price but would be depressed in the longer run because of the reduced incentive to replace depleted reserves. This adverse effect will partly offset the benefits of lower import bills.

27

OPEC and Future Oil Supplies

The Organization of Petroleum Exporting Countries came of age in the 1970s. In its prime, it presided over the dramatic restructuring of the international petroleum market, but it now faces the challenge of adjusting to a new set of market realities. This challenge comes at a time when the organization is facing severe internal strains. The victory of Ayatollah Khomeini over the shah of Iran not only precipitated the oil price shock of 1979 but also created profound tensions among the leading members of OPEC, notably Iran, Iraq, and Saudi Arabia.

In an earlier era, OPEC policy was set by Saudi Arabia and Iran together, and sometimes even by Saudi Arabia alone. But by the early 1980s it was clear that the political strains within OPEC were becoming a problem. To cut price or output would mean a loss of income for some or all OPEC states. For most members this represented a genuine economic threat and a loss of power in the international arena. For some regimes, it might prove to be fatal. So the stakes were large as OPEC met in 1982 for the first time to consider a price cut and to confront the possibility of production quotas for individual member states.

The problem with agreeing on a price cut was to determine one that would be sufficiently large to stimulate demand while not causing too great a revenue loss in the short run. Some countries in OPEC were better positioned, because of small populations and substantial financial reserves, to forgo present income in anticipation of a later recovery of demand.[23] Others had very pressing needs for current income and would have preferred to defend high prices by having the countries with large financial reserves make sizable cuts in production. Adding to the complexity was the fact that no one could know for sure how great the price cut would have to be, how much the market would respond, or how long the response would take. OPEC would be entering a realm of great uncertainty.

If price cuts were politically difficult for OPEC to manage, production quotas were even more complicated to negotiate. Several possible formulas might have been thought fair. Each OPEC member might, for example, be expected to make identical percentage cuts from current production. This, however, would favor those who happened to be pro-

23. Richard P. Mattione, *OPEC's Investments and the International Financial System* (Brookings Institution, forthcoming).

ducing near capacity at the time of the decision, while discriminating against those who might be temporarily producing at lower than normal rates. Thus an alternative might be to use historical market shares to set quotas, which would work in favor of countries like Saudi Arabia and Iran but would hurt Iraq and some of the newer members. The oil-producing states with large populations, such as Nigeria, Algeria, Indonesia, and Iran, held that market share should be related to economic need and level of development. This was an argument intended to put pressure on Saudi Arabia alone for any substantial cuts in output.

Given the variety of interests within OPEC and the competing formulations on production sharing, it was inevitable that power and politics would help determine the outcome of the debate. In March 1983 OPEC reached agreement on market shares based on a production ceiling of 17.5 million barrels per day (bpd). Saudi Arabia declined to accept a set figure for production, but nonetheless seemed willing to act as a "swing producer," reducing output below 5 million bpd if total demand for OPEC oil fell under 17.5 million bpd. According to the Saudis, however, if demand were to exceed 17.5 million bpd, the other OPEC members should stick to their quotas while Saudi Arabia increased production. No one else in OPEC agreed with this interpretation, least of all Iran. In short, OPEC was still far from having found an acceptable set of guidelines for controlling production, and still had no means to discipline the behavior of any member state that might exceed its quotas. Ultimately, it was the threat of breaking the price structure or of military and diplomatic pressure that held the organization together.

Iran, Iraq, and Saudi Arabia: Triangular Politics within OPEC

OPEC's three giants, Iran, Iraq, and Saudi Arabia, were capable together of producing about 18 million bpd as of mid-1984. This would be about equivalent to all the oil that OPEC exported at that time. During 1983 and 1984, each of these three countries produced at 50 percent or less of its maximum sustainable capacity. In the case of Iran this was in large measure due to the dislocations caused by the revolution, which resulted in poor oil-field maintenance as well as some reluctance on the part of consumers to become dependent on an unsure source of oil. In these circumstances Iran was able to export about 2.0 million to 2.5 million bpd, earning some $20 billion annually. Over time, with new investment in oil-field development and a more benign regional climate, Iran could expect to increase exports significantly, although probably not to the level of 6 million bpd that was reached under the shah.[24]

Iraq has found itself in a uniquely awkward position as an oil producer. Throughout most of the 1960s, disputes with international oil companies kept Iraq from expanding its output. Finally, in the 1970s Iraqi oil output began to take off. It was widely believed that Iraq possessed enormous undeveloped reserves of oil, especially in the south. By 1979 exports were more than 3 million bpd. Then, in 1980, Iraq invaded Iran, and in reply the Iranian navy destroyed Iraq's offshore loading platforms in the Persian Gulf. This immediately deprived Iraq of its single most important means of exporting oil. The regime in Baghdad was left heavily dependent on two pipelines, one passing through Turkey that could carry nearly 1 million bpd once it reached full capacity, and one crossing Syria with a capacity of about 800,000 bpd. Unfortunately for Iraq, Syria aligned itself with Iran in the war and decided in 1982 to close the Iraqi pipeline passing through its territory. Throughout 1983 Iraq struggled to sell as much as 1 million bpd, earning less than $10 billion for the year.[25]

Iraq desperately wanted to be able to increase output, which was below its designated share within the OPEC formula, but it was not physically able to do so. Saudi Arabia reportedly marketed some of its own oil under the Iraqi quota, with the proceeds going to Baghdad. Still, the realization that political developments could constrain the Iraqi economy because of the vulnerability of the oil-export routes led the Iraqis to look for alternative outlets. Geography made this search difficult.[26] One attractive possibility was to build a pipeline south through Saudi Arabia to link up with Petroline, the east-west Saudi pipeline that carries crude oil and natural gas liquids to the Red Sea port of Yanbu. Since the Saudis rarely used the line to full capacity, it was thought that the Iraqis might borrow space in it until a completely new line for Iraqi oil to Yanbu could be completed. If all worked well, this could mean that by 1986 the Iraqis might increase exports via the Red Sea by some 500,000 barrels per day. But this would also increase Baghdad's dependence on conservative Saudi Arabia. And it

24. Shaul Bakhash, *The Politics of Oil and Revolution in Iran* (Brookings Institution, 1982).

25. The pipeline through Turkey is probably Iraq's most secure outlet for oil. Iran is heavily dependent on trade with Turkey and is reluctant to cut the pipeline from Iraq to Turkey since Turkey might retaliate by closing the border with Iran. This argues in favor of trying to expand the capacity of the line through Turkey and has the indirect benefit of helping the Turkish economy.

26. Christine Moss Helms, *Iraq: Eastern Flank of the Arab World* (Brookings Institution, 1984).

was not at all clear that the Saudis genuinely welcomed the idea of such a close link to Iraq.[27]

As an alternative to a pipeline through Saudi territory, the Iraqis may proceed with plans for a line through Jordan to Aqaba. Jordan would welcome the project and would benefit materially from it. And the line would add new export capacity, perhaps as much as 1 million bpd, that would be available even if shipping in the Gulf were to be disrupted. From Iraq's standpoint, it would reduce dependency on the Saudis, which would give Iraq a freer hand within OPEC. However, the pipeline to Aqaba would place Iraqi oil within very easy striking range of Israel, which would presumably have some deterrent effect on Iraqi actions in the event of future Arab-Israeli conflicts. In mid-1984 this project received official U.S. support in the form of Export-Import Bank credits and an American firm, Bechtel, was preparing to oversee the construction.

A third alternative for Iraq would be to increase the capacity of the pipeline through Turkey. Turkey was relatively immune to Iranian pressures, shared an interest in keeping the Kurdish areas along the common border quiet, and had proved to be a good partner in the oil business. There was also the possibility of tapping an existing Turkish line that had spare capacity.[28]

While it remained impossible to be certain as of mid-1984 which pipeline the Iraqis would ultimately develop, it seemed clear that some exports through one of these routes would be part of the Middle East scene by 1986. The Iran-Iraq war provided strong incentives for Baghdad to develop these new export routes.

Saudi Arabia is, of course, the most important member of OPEC because of its immense reserves and its huge production capacity.[29] In the early 1980s Saudi Arabia was able to sustain output at more than 10 million bpd, making it the second largest oil producer in the world at that time. By 1983 Saudi exports stood at less than half the maximum capacity, but still the kingdom was the third largest producer of oil in the world after the Soviet Union (the largest) and the United States. With 5 million bpd of

---

27. In the event of a disruption of oil shipments from the Gulf, Saudi Arabia would want to be in a position to use the full capacity of Petroline. Politically, it might be difficult in such an emergency to deny the Iraqis access. Furthermore, the Saudis have reportedly urged the Iraqis to think in terms of a pipeline all the way to Yanbu, not just a spur linking up to Petroline.

28. "Will the West Build Baghdad's Life Lines?" *Middle East Economic Digest*, vol. 28 (March 2–8, 1984), pp. 24–29.

29. William B. Quandt, *Saudi Arabia's Oil Policy* (Brookings Institution, 1982).

spare capacity, Saudi Arabia is the only country in the world that could flood the market and bring the current price structure down. It also has the capacity to make up for the loss of all Iranian and Iraqi oil in the unlikely event that the Gulf war were to escalate and disrupt the oil exports of the two belligerents. In addition, Saudi Arabia has considerable influence over the oil policies of other members of the Gulf Cooperation Council, especially the large producers such as Kuwait and the United Arab Emirates; acting in concert, these countries could control output of 15 million bpd.

Despite its apparent strength as an oil power, however, Saudi Arabia is vulnerable to many pressures. It has a small population, by regional standards is militarily weak, and faces potential threats from every direction.[30] Its wealth and resources are coveted. In these circumstances, the Saudis have developed a cautious approach to the problems around them.[31] They shun confrontation and seek consensus. They have taken out multiple insurance policies. They have sought help from the United States, but have not wanted a close embrace. They have extended enormous economic assistance to Iraq during its war with Iran, but have not wanted to become embroiled in the fighting and have sought to keep some lines open to Iran.

What will this triangular rivalry among Iran, Iraq, and Saudi Arabia mean for OPEC in the second half of the 1980s? First, genuine cooperation on oil policy among the three giants of OPEC is unlikely, and thus OPEC is unlikely to be in a strong position to raise prices. Iran and Iraq will remain locked in a bitter conflict, even if the war comes to an end. The potential will remain, despite temporary truces, for a resumption of warfare at a later date, and oil facilities and shipping could be at risk. Even without open warfare, Iraq and Iran will be seeking greater shares in the oil market. They will probably be on the side of the price hawks, since both countries have heavy financial burdens and the war has caused great damage and dislocation that will be costly to repair. One interesting possibility is that both Iran and Iraq will at some point seek to increase output at the expense of the Saudis, and both will push for somewhat higher prices than the Saudis will want to support.

Iraq may be tempted to press the Saudis for a share of their market, but on balance it seems that Iraq will remain dependent on Saudi Arabia for political and economic support. This leaves Iran as the more serious rival

30. Thomas L. McNaugher, *Arms and Oil: U.S. Military Security Policy toward the Persian Gulf* (Brookings Institution, forthcoming).

31. William B. Quandt, *Saudi Arabia in the 1980s: Foreign Policy, Security, and Oil* (Brookings Institution, 1981).

of Saudi Arabia. On ideological grounds, the Islamic Republic in Iran is opposed to the conservative monarchy across the Gulf. Within OPEC, Iran will use the argument, which is attractive to all other members, that Saudi Arabia should keep production low when demand for oil is low, and that others should be allowed to expand production before Saudi Arabia does so. How this will sort itself out is difficult to foresee, but the Iranian case will probably win considerable support. This will put the Saudis in the position of either breaking the OPEC price structure by keeping output high, which is the ultimate threat the Saudis can make to the other members, or acquiescing to their demands, at least to some extent, and keeping production in the 4 million to 5 million bpd range. This is less than many Saudis would like, but it is certainly enough to carry forward an impressive development program.

The most serious economic problem for the Saudis will come in the unlikely event that the Iran-Iraq war comes to a sudden end before demand for oil has recovered from its 1983–84 low. Iran and Iraq would quickly try to increase their oil exports. Within a short period, some 2 million to 3 million bpd of oil would be on the market. Great pressure would be placed on the Saudis to make most of the cuts necessary to open the way for this amount of oil. In such circumstances, Saudi production could be forced down as low as 2 million bpd, well below the level that any Saudi would consider adequate to meet the country's need for revenue. Other members of OPEC would find it difficult to make more than token cuts. This would set the stage for competitive price cutting and, in the worst circumstances, an uncontrolled drop in price toward $20 or so per barrel. All OPEC members would be worse off in the short run, with the possible exception of Iraq, which with 3 million bpd of exports at $20 would be earning twice as much as in 1983. But Iraq would have to consider that in such circumstances financial aid from the Gulf Cooperation Council would almost certainly be terminated, so that even Iraq might not be better off on balance.

In brief, the strains that have always existed within OPEC are likely to be exacerbated in the remaining years of the 1980s for several reasons:

— The oil market will remain comparatively weak, obliging most OPEC members to hold production well below desired levels.

— Prices are unlikely to rise appreciably, and a price increase will not boost national incomes as in previous years.

— Iran will remain a revolutionary and disruptive force in the Gulf, even though a tendency toward moderation can eventually be expected.

— Iran and Iraq will both remain intent on increasing exports of oil, especially if the war comes to an end. This will add a strong element of competition to OPEC deliberations.

— Saudi Arabia could come under severe pressure to make cuts in production that would fall well below the minimum income requirements of the country. In such circumstances, the Saudis would probably argue for a reduction in prices rather than accept the full burden of making cuts.

— No stable, equitable production-sharing formula is likely to be devised within OPEC.

— Political rivalries within the organization will preclude the development of effective policing mechanisms to be used against members who shave prices or exceed quotas.

If this picture of OPEC in the future is correct, the price of oil will be determined less by concrete decisions by its members than by the rate of economic recovery in the industrialized world, the course of the Iran-Iraq war, and the possibility of international developments that may cause significant supply interruptions. In short, OPEC will be lucky to be able to prevent severe price declines and will depend on external factors, not deliberate decisions, to produce circumstances in which prices might actually be increased.

Possibilities of Supply Disruptions

Sometime in the coming years it is almost certain that oil supplies will be disrupted as a result of war, revolution, or political action. Most likely, any dangerous disruption will involve the Persian Gulf, since only there are the quantities of oil so large that a major price shock could result. In addition, the Gulf region is rife with conflict and potential for political change. Even the threat of large-scale disruption in this sensitive area can have important short-term effects on prices.

In contrast to the 1970s, however, disruptions in the 1980s need not produce large and persistent price increases. As indicated earlier, for any magnitude and duration of a disruption, the outcome will depend on underlying market conditions and the decisions made by the private sector at the onset of the crisis. Sound public policies can also help minimize the price shock accompanying any disruption, but sound policies will require unusually sophisticated understanding of the political and strategic forces that will have brought about the loss of oil.

War in the Persian Gulf. The most credible scenario for significant oil disruptions in the 1980s involves Iran and Iraq, two sizable oil producers with considerable military capability and the demonstrated ability to damage one another's oil installations and tanker traffic in the Gulf. The Iran-Iraq war began in September 1980 and in its first years demonstrated how easily oil could be removed from the market as a result of the war. The Iranian navy successfully attacked the Iraqi offshore loading terminals in the Gulf, making it impossible for Iraq to market some 2 million bpd of oil. Iran's ties to Syria resulted in a decision by the Damascus regime to close the Iraqi pipeline across Syrian territory in 1982.

In periods of stalemate and prospective negotiations, it is reasonable to expect that both Iran and Iraq might refrain from hitting each other's oil facilities, but a negotiated end to the war has been elusive. Indeed, the two military powers of the Gulf seem to be set for a prolonged conflict. There may be truces along the way, but the underlying conflict between Tehran and Baghdad, grounded in clashes over ideology and national interests, is likely to persist well into the future. To break the stalemate, the Iraqis began to talk of escalation in late 1983. The war of attrition seemed to be working against them, and their receipt of new, long-range Super Etendard jets and Exocet missiles from the French led to much speculation that the Iraqis might try to destroy the Iranian export terminals at Kharg or to strike at oil tankers calling at Iranian ports. Such attacks were in fact launched in March 1984, causing a tremor—but little more—in the oil markets and leading to increased insurance costs for oil tankers entering the Gulf.

Assuming that the Iraqis could seriously disrupt Iranian exports, how might the Iranians respond? The regime in Tehran has explicitly threatened to close by force the Strait of Hormuz if Iraq destroys Kharg or drives away customers. Iran could threaten tanker traffic in the Strait and the rest of the Gulf by mining, air and artillery attacks, or naval blockade. Any such move would frighten some oil customers away from the Gulf, and insurance rates for shipping would certainly rise, as they did in May 1984 after a number of Iraqi and Iranian air strikes against tankers.

But do the Iranians really have the ability to close the Gulf and keep it closed? The answer depends largely on the American response. It would not be difficult for the U.S. fleet in the Arabian Sea, joined perhaps by French and British ships, fairly quickly to reopen the Strait and provide safe passage for tankers. There might be a period of some uncertainty during which tankers would stay away, and there would doubtless be costs in any such operation. But the Soviets would not be likely to intervene and the Iranians could not long resist.

A more dangerous scenario, one which could be triggered even without a successful Iraqi strike against Kharg, would involve Iranian threats to

Kuwait and Saudi Arabia, the primary financial supporters of Iraq. The objective would presumably be to force these two countries to end their backing for Iraq. One tactic the Iranians could use would be terrorist attacks against oil installations. Iran's supporters in Lebanon and the Gulf have proved adept at the use of car bombs, several of which went off in Kuwait in December 1983. There was no serious damage, but the oil installations are vulnerable.

A second Iranian tactic might be to launch air strikes against oil facilities in Kuwait and Saudi Arabia or against tankers calling at their ports. Both countries, but especially Kuwait, could be vulnerable to such attacks, and considerable damage could be done. But air strikes, to be successful, would have to be carried out on a very large scale and over a prolonged period. No single attack could be expected to do much damage, and the Saudis have substantial means of air defense against Iran. Not only do they have an early warning system, but they have an air force that with 150 fighter planes should be able to provide some defense against the small Iranian force. More important, the Saudis know that they could ask for, and expect to receive, American help to provide an effective air defense against Iran in an emergency. To be effective, however, direct American military assistance would have to use the airbase at Dhahran, and the Saudis would be reluctant to offer such access in any circumstances other than a major, sustained Iranian threat.

A more serious, long-term danger to oil supplies in the Gulf would result from a major Iranian penetration of Iraqi defense lines in the south. The Iraqis appear to be strong enough to prevent such a breakthrough, but if the Iranians are successful it would mean a sudden shift in the balance of power. Iran would be in a commanding position, not only to pressure the regime in Baghdad, but also to intimidate Kuwait and to threaten the Saudis. The specter of masses of Iranian troops poised on the Kuwaiti border is a nightmare to all Gulf Arab leaders. This is one military scenario in which outside support would be essential but of questionable effectiveness. In those circumstances, many Gulf leaders would be tempted to accommodate the revolutionary regime in Iran. This could give Iran much more say over the decisions on oil production made by all of the Gulf countries. Oil would, of course, still flow to the West, but a dominant Iran might be able to curb some of the competitive pressures within OPEC, making it more able to administer prices than would otherwise be the case, especially later in the 1980s, when the market can be expected to be tighter than it is today.

If this analysis is sound, the Iran-Iraq war may result in some further small-scale disruptions of oil supply, but the threat of a very large scale disruption could be dealt with effectively by American military action. The analysis provides no reason to be sanguine, but it does direct attention to the remedy for a major disruption if one were to occur.

While American public opinion, allies, and local sensitivities might inhibit U.S. military action in some parts of the world, the Gulf is one area where a president could order military action in defense of oil supplies and count on widespread support at home, in the region, and in virtually every oil-importing country. At worst, there might be a few months of uncertainty about the resumption of oil flows. Actions such as Saudi stockpiling of oil on ships at sea could give an added margin of safety. In any case, supplies of oil from the Gulf would most likely be restored soon to predisruption levels. The long-term danger does not lie so much in the temporary disruptions that may result from the Iran-Iraq war as from the possibility that a victorious Iran might be able to dictate oil policy for much of the Gulf region, thereby adding strength to OPEC.

Instability in Saudi Arabia. If the Iran-Iraq war seems to pose real, but manageable, threats to the world oil market, the same could not be said if Saudi oil production were placed in jeopardy, either as an outgrowth of a regional conflict or for internal reasons. Saudi reserves, production capacity, and current production are simply too large for the world to do without for very long. The case of possible external aggression against Saudi Arabia, while dangerous, is probably easier to handle than an internal upheaval that might spread to the oil fields.

The Iranian revolution has never seemed to be a good model for what might happen in Saudi Arabia. Nonetheless, it is worth recalling what happened to Iranian oil production after that revolution. Production fell dramatically over a period of several months as a result of strikes in the oil fields. Foreign workers fled or were expelled, which greatly reduced the productive capacity of the fields because of poor maintenance. Moreover, in the early phase of the revolution, production was deliberately held to low levels as an ideological debate over oil policy took place.[32] Were the same developments to occur in Saudi Arabia, several million barrels per day of oil might be lost to the world market, and, perhaps more important, doubts would immediately arise about the availability of Saudi reserves in the future. The major safety valve in the world oil market would be in question. Consumers could be expected to scramble for oil, bidding prices up significantly.

32. Bakhash, *Politics of Oil*, pp. 1–12.

37

Saudi Arabia certainly faces internal strains as a result of the rapid pace of social and economic development coupled with the relative lack of change in the political system. But the kingdom does not seem to be heading toward a disruptive internal upheaval. If and when internal political changes take place, they are most likely to involve the burgeoning class of technocrats and the military. These may be nationalistic groups but not ones necessarily imbued with great ideological zeal.

Even if it is difficult to imagine drastic internal upheaval in Saudi Arabia, the consequences of such a development would be so serious for the world that some thought must be given to the possibility. One problem for outsiders would be to decide whether and when to intervene if internal Saudi developments threatened to create serious instability. Would intervention be warranted by a palace coup? By persistent strikes and terrorist bombings in the eastern province? By reform-minded soldiers seizing power? Ambiguity would necessarily surround such events, making a decisive response by the United States much more difficult. Could the president count on support for sending troops to the Saudi oil fields without a clear-cut threat from a hostile government to withhold the oil? Even then, would not some argue that the United States should let the dust settle before using force? After all, one can imagine a situation where precipitous action might be worse than none at all.

It does little good to devise precise guidelines for dealing with an upheaval in Saudi Arabia. Fortunately, the prospects in the near term for such an event do not seem especially great, and the most probable types of change would have relatively little impact on oil policy. One Saudi prince is just about as likely as any other to keep the oil flowing on about the same terms, at least so long as a relatively soft market persists. And a nonroyal regime would also need large infusions of revenue and would seek to market oil.

Only in the event of the return of very tight markets will the world need to be worried about the most probable kinds of changes that may occur in Saudi Arabia. This suggests that the period of the early 1990s could be a particularly sensitive one as the likely tightening of the market may combine with internal pressures in Saudi Arabia to produce an explosive result.

Arabs versus Israel. The Arab-Israeli conflict, which helped to trigger the first oil shock in 1973, must be considered in any analysis of oil supply in the 1980s. Before the Arabs can use the oil weapon again in their struggle against Israel, three conditions must be met: a high degree of solidarity among the Arab oil producers; comparatively tight markets; and a catalyst to precipitate collective action. It is not hard to imagine a Syrian-

Israeli war in the latter half of the 1980s. One can even assume that such a war would cause Arab producers to take collective action against the United States. But such action would not have anything like the impact it had in 1973.

Much has changed in the past decade. No one now would be particularly worried by a selective embargo of oil shipments to the United States. It was not the embargo that hurt in 1973, even though it was the most visible action by the Arab oil producers. The real problem arose because of coordinated production cuts by exporters and rapid stock building by consumers. In a tight market, a few key producers could make a major difference if they were to announce cuts in output, but they would have to be prepared to accept the adverse long-term consequences for their export earnings. In relatively soft markets, this action would simply provide other oil producers with the opportunity to increase sales.

As argued earlier, economic conditions are such that there seems to be relatively little reason to fear a repetition of 1973 in the near future. The Lebanon war of 1982 was not accompanied by any such threats. In addition, Arab solidarity on the Palestinian issue has certainly been weakened by the events of 1982–83, so that some Arab regimes secretly might be quite pleased to see the Syrians engaged in a war with Israel even if the public reaction of these regimes would be one of solidarity with Syria. The net effect would likely be that collective action by Arab oil producers would not amount to much. The caveat, as always, must be that the return to a very tight market would provide any large oil producer with some access to the oil weapon. In short, one cannot be indifferent to the dangers to oil supplies in the wake of a major Arab-Israeli war, especially if this occurs in the late 1980s or early 1990s, but the chance of a repeat of 1973 anytime in the next few years is very remote.

Serious and Less-Serious Disruptions

When the next oil supply disruption occurs, officials and consumers will have to act on judgments about the kind of disruption that has taken place. We have argued that because of structural changes, future disruptions will take a different course than they have in the past. Nevertheless, the problem for analysts, policymakers, and consumers alike is that there will be a great deal of uncertainty surrounding any significant disruption. Would an Iranian attack on Saudi oil facilities be the opening phase of a prolonged and destructive war, or would it be more in the nature of a warning shot? Would demonstrations in the eastern province of Saudi

Arabia portend a mass uprising, or would they remain limited? Would an Iraqi airstrike on Kharg Island do permanent damage, or could repairs be made easily? Private consumers are likely to pay considerable attention to the words and actions of governments at such times. In addition to diplomatic initiatives and military deployments, governments will give signals by how they choose to use their public stockpiles. The U.S. government has announced the intention to use its petroleum reserve early in a disruption to dispel panic behavior, yet to use the reserve a national emergency must be declared. The announcement would add to the sense of crisis.

With the considerable spare capacity and flexibility in consumption that are likely to prevail for several years, the primary question posed by a disruption is whether any fundamental structural change in the market has occurred. In the case of the Iranian revolution, such change did take place, both in the nature of the Iranian regime and in the likely ultimate recoverable capacity of Iranian oil reserves. As a result, some price increase was probably inevitable, although not necessarily the doubling that occurred.

In the case of large disruptions, on the order of 5 million bpd over a period of several months, a number of specific decisions would have to be confronted by oil-importing countries. For the industrialized nations that are members of the International Energy Agency, any loss that reduces consumption beyond 7 percent could trigger collective actions such as sharing, stock draws, and demand-restraint measures. These steps would not be entirely automatic, nor are they specified well enough to assure success in stabilizing the market.

Beyond collective economic and oil-management decisions, the United States would have to confront the question of whether to use military force and in what way. Some actions, such as reopening the Strait of Hormuz if it were mined, should not require prolonged debate and should not pose unacceptable risks; others, such as intervening in Saudi Arabia in the midst of domestic upheaval there, would be extremely difficult to assess, and alternatives to the use of military force might well be preferred. Last, the unlikely event of a Soviet threat to the Gulf region's oil would raise the question of whether it would be possible for the United States to meet a Soviet military challenge in the Gulf with conventional means.[33]

The dangers and uncertainties accompanying any sizable disruption of oil supplies are so great that it is worth major diplomatic efforts to forestall their occurrence. This conclusion does little to help shape the contours of a

33. McNaugher, *Arms and Oil*; and Joshua M. Epstein, "Soviet Vulnerabilities in Iran and the RDF Deterrent," *International Security*, vol. 6 (Fall 1981), pp. 126–58 (Brookings Reprint 401).

policy, but it does suggest that the traditional American concerns for Middle East stability, for a resolution of the Arab-Israeli conflict, and for containment of Soviet influence in the Gulf region are all consistent with sound considerations of energy security. The challenge is to devise policies that advance this set of interests, recognizing that diplomacy will always have to be buttressed by credible military capability.

The International Setting and Energy Security

For the rest of the century the ideal international environment for oil importers would be stability in the Middle East, détente in U.S.-Soviet relations, and a high degree of cooperation among the major industrialized countries. In such a setting, the risks of an oil disruption would be low, the military dangers accompanying any such crisis would be minimal, and collective remedial action would be available if needed. Market forces could generally be relied upon to allocate energy supplies efficiently.

To describe such a benign environment is to suggest why energy security is likely to remain a problem in the 1980s and beyond: not because of a shortage of crude oil or of productive capacity, but because, as in the past, political and military developments can get in the way of the market. In particular, it is the combination of instability in the Middle East, U.S.-Soviet rivalry, and competition within the Western alliance that creates the potential for oil shocks in the future. Is any of this likely to change significantly in coming years?

The Soviet Union and Energy Security

The Soviet Union is the second largest producer of energy in the world today, after the United States, and the largest producer of oil. Soviet oil and gas exports in 1982 equaled 4 million bpd of oil; the exports earned for Moscow about $18 billion in hard currency, one half of it from the West. Energy exports to Eastern Europe, formerly provided by the Soviets at heavily subsidized prices, still constitute an important element of Moscow's influence over the other members of the Warsaw Pact.

In the late 1970s it was widely reported, especially by the Central Intelligence Agency, that Soviet oil output would begin to drop. In fact, the Soviets have been able to keep oil production at about 12 million bpd, but the cost of doing so has risen. There seems to be little prospect for

41

expanding the oil sector of the economy, and sometime in the late 1980s output may begin to drop off.

As their oil sector stagnates, the Soviets have the luxury of choosing among several strategies for meeting their energy needs without becoming dependent on imports. First, and perhaps least likely, would be significant reforms in the economy that would result in greater efficiency of energy use. Second, and already the leading candidate, is the expansion of natural gas production. Reserves are ample, the infrastructure investments are being made, and an export market and credit sources are available in Western Europe to defray the costs of development. Gas can also displace oil in the domestic market to ensure that oil will remain available for export. Third, the Soviets could make greater use of their large reserves of coal.

Looking at these trends some Western analysts have seen two dangers for energy security. First, as Soviet oil output declines, might not the Soviets seek access to oil supplies in the Persian Gulf or North Africa, both for themselves and for Eastern Europe? Since hard-currency purchases are virtually precluded, the Soviets would have to secure the oil through barter deals—weapons for oil—or through diplomatic and military pressures. In extreme circumstances, some have argued, the Soviets would be tempted to intervene militarily in Iran to seize the oil fields there and to control the entire Gulf region.

The weakness in this analysis is that the overall energy balance in the Soviet Union is likely to remain positive well into the future.[34] Desperate actions such as invading the Gulf would therefore not grow out of economic necessity. If the Soviets were ever to see a strategic advantage in controlling Gulf oil, it would be not because of their own needs but because of the leverage it would give them over Europe and Japan. The offsetting consideration would obviously be the risk of global war that would lie behind any Soviet threat to the Gulf. Deterrence of an overt military move by the Soviets toward the Gulf will depend on both the regional balance of forces, which favors the Soviets—but not overwhelmingly—and the global balance of power that would come into play in any superpower confrontation in the Gulf.

For reasons of broad international strategy, but not of economic necessity, the Soviets will no doubt seek to expand their influence in the Gulf.

34. Ed A. Hewett, *Energy, Economics, and Foreign Policy in the Soviet Union* (Brookings Institution, 1984).

To do this they will most likely provide arms and diplomatic support to friendly regimes and take some covert actions in order to seize opportunities when they arise, and on occasion they will probably try to create such openings. For the West, this is a familiar challenge, even if the appropriate response is often difficult to agree upon.

A second Soviet incentive rooted in economic realities is to keep foreign exchange earnings at high levels. This means exporting arms to those countries that can pay for them, largely in the Middle East, and selling oil and natural gas to Western Europe. The impulse to sell arms means that conflicts in the Middle East will be particularly lethal, as both superpowers bestow on their friends and clients the latest in deadly technology. The most dangerous conflicts are likely to be the one between Iran and Iraq, which has a direct bearing on oil supplies, and between Syria and Israel, which is less closely related to energy security.

Soviet exports of energy to Western Europe have been a controversial issue within the Western alliance. From the standpoint of economics, it makes sense to encourage the Soviets to produce oil and gas for the world market. This increases the amount of energy that is not subject to OPEC decisions, puts downward pressure on prices, and promotes price stability. To gain customers, the Soviets are obliged to sell at or below market prices. To retain customers, especially when so many alternative sources are available, the Soviets must cultivate their reputation for reliability.

Two arguments have been raised against the export of Soviet energy to Europe. One says it subsidizes the Soviet economy, and the other says it reduces Western security. The economic objections are that the credit arrangements for financing the gas pipeline to Europe and the pricing formula for the gas are too liberal and that the Soviets did not pay adequately for their access to the advanced technology to build the compressors for the pipeline. By 1983 the debate over the economic issues had essentially been settled, with Europeans asserting that the gas deal made economic sense for them and that the transfer of technology was minimal. The American government acquiesced but continued to worry about the security argument, which is that the deal created excessive European "dependence" on Soviet energy supplies. Some Europeans felt that the arrangement might be questionable on economic grounds if the price of oil were to fall, but few seemed worried by the security argument.

The dependency argument is a familiar one, applicable to the Middle East as well as the Soviet Union. For the Europeans and Japanese, dependence on some foreign sources for energy is inevitable. The point is to

43

keep dependency from creating vulnerability.[35] This is best accomplished by diversifying sources of imports, developing sizable stockpiles, creating fuel-switching capability, and promoting energy efficiency by market pricing of fuels. To varying degrees, most of the European countries have taken steps in these directions.

In summary, it does not appear that European dependence on Soviet energy supplies in the coming years will create a significant energy-security problem. In fact, Soviet supplies of energy to Europe may help to keep world oil prices lower than they would otherwise be. In addition, development of Soviet gas fields will provide the Soviets with an alternative to Middle Eastern oil for themselves and their allies, thus reducing whatever temptation they might otherwise have to use military power to gain access to oil reserves abroad.

The Course of U.S.-Soviet Relations

Even at the height of the era of U.S.-Soviet détente in the early 1970s, good relations between the superpowers did not prevent intense competition in the third world.[36] Consequently it strains the imagination to believe that the United States and the Soviet Union, at a time of poor relations in the 1980s, will be able to work together to promote stability in sensitive areas like the Middle East.

Nonetheless, proposals are occasionally advanced, especially by the Soviets, that call for joint efforts to ensure the free flow of oil from the Persian Gulf. This was notably the case with the Portugalov statement in 1980.[37] It was widely seen as a bid by the Soviets to enlist European support for a Soviet role in any security arrangements for the Gulf. No one in the Gulf region showed much interest, and the initiative led nowhere.

While explicit agreements between Moscow and Washington concerning the Gulf are extremely unlikely, both powers do have an interest in

---

35. David A. Deese and Joseph S. Nye, eds., *Energy and Security* (Ballinger, 1981), chap. 1.

36. U.S. and Soviet competitive intervention on opposing sides of the 1973 Arab-Israeli war is a vivid example of the limits of détente. See the study by Raymond L. Garthoff, *Détente and Confrontation: American-Soviet Relations from Nixon to Reagan* (Brookings Institution, forthcoming).

37. Nickolay Portugalov commentary, "An Alternative," Tass, February 29, 1980, in "Portugalov Calls for European Conference on Persian Gulf," Foreign Broadcast Information Service, *Daily Report: Soviet Union*, March 3, 1980, pp. G1–G2. The commentary urged Europeans to pay attention to Leonid Brezhnev's February 22, 1980, speech on the Persian Gulf, which was the Soviet answer to the "Carter Doctrine."

making sure that the other understands its vital interests. This requires some degree of dialogue and some tacit understandings.

At least since the enunciation of the "Carter Doctrine" in January 1980, the United States has been on record saying that it would be prepared to use force to defend its interests in the Gulf. President Reagan added that the United States might be prepared to intervene militarily to prevent the overthrow of the Saudi regime. This still leaves many questions unanswered about how the United States would behave in a variety of circumstances, and it begs the question of whether it would be able to carry out its declared policies.

By contrast, the Soviet Union has stated its interests in the Gulf region in more defensive terms. The Soviets readily note that this is a region close to their own borders. On occasion they mention the 1921 treaty with Iran that permits the USSR to send troops into that country in the event of a collapse of law and order. Those with some memory of the past will recall Soviet efforts after World War II to establish autonomous republics in Azerbaijan and Kurdistan in northern Iran. In the Iran-Iraq war, they have not intervened directly on either side, although they have "tilted" on occasion to help prevent the defeat of either power.

The Soviets have not sought to use their substantial military power on Iran's northern border to threaten the Gulf. The Soviets do not refer to a need for access to ports in the Persian Gulf or the Arabian Sea. This is much more a fixation of Western analysts and involves a misreading of Soviet history. When the Soviets referred to their interests in waters to their south in the Nazi-Soviet talks on a nonaggression pact in 1940, it was the Black Sea and the Dardanelles that concerned them, not the Gulf or the Arabian Sea.[38] From a strictly military standpoint, a port in the Gulf would be terribly vulnerable in the event of hostilities because of the choke-point of the Strait of Hormuz, through which shipping would be obliged to pass. A port on the Arabian Sea would make more strategic sense but is not a likely prospect or a high priority.[39]

The Soviets seem to recognize that the oil of the Gulf is a vital U.S. national interest that cannot be lightly challenged, and they have expressed some concern that instability in the area could provide a pretext for American military intervention. Similarly, the Americans sense that Afghanistan

38. Arnold L. Horelick, "Soviet Policy in the Middle East, pt. 1: Policy from 1955 to 1969," in Paul Y. Hammond and Sidney S. Alexander, eds., *Political Dynamics in the Middle East* (American Elsevier, 1972), pp. 560–61.

39. Michael K. MccGwire, *Soviet Military Objectives* (Brookings Institution, forthcoming).

and northern Iran are areas of Soviet military supremacy that would be difficult to challenge by conventional means. This has led some to anticipate that a tacit agreement could emerge between the superpowers that would essentially divide the Gulf region into a northern Soviet zone and a southern U.S. zone. From an energy standpoint, the West would be the big winner. But such an arrangement ignores the fluidity of the regional political environment, the difficulty for either superpower of maintaining control in this area, and the significant military advantage that the Soviets would gain if they did succeed in occupying northern Iran.

The Central Command: Defense for the Gulf?

If a U.S.-Soviet understanding concerning the Gulf seems exceedingly unlikely, and spheres of influence look shaky at best, is the United States in a position to rely on some combination of diplomacy and military power to deter Soviet moves into this vital area? Beginning in 1979 the United States sought to strengthen its ability to project military force into the Gulf region. The problems were formidable. The lack of reliable forward bases, the logistical problems involved in maintaining a force halfway around the world, the inhospitable climate, the inappropriateness of much equipment and doctrine designed for the European theater—all of this had to be confronted by policymakers. In addition, constraints on budget and personnel limited how much could be done and how quickly.

Nonetheless, progress was made in ensuing years. Some equipment and supplies were positioned closer to the region; access to facilities in Oman, Somalia, Kenya, and Egypt was negotiated; exercises were carried out to familiarize U.S. troops with the harsh environment in which they might have to operate; mobility was improved, especially through sealift; and a new organization was created called the Central Command.

Despite these changes, it was not at all clear how the United States would be able to react if the Soviets were to threaten the Gulf. By most accounts, Iran would be the strategic prize, with the oil fields of Saudi Arabia the most valuable asset. In view of the overwhelming Soviet advantage in northern Iran in the event of war, some strategists were inclined to limit U.S. efforts to defense of the oil-producing areas themselves, essentially ceding the north to the Soviets. Others felt that this would give the Soviets an overwhelmingly powerful position and would make control over the south problematical as the Soviets consolidated their position.

This latter view argued for developing an ability to blunt any initial

Soviet move into northern Iran, relying on air power and troops trained for mountain warfare to stop the Soviets as far north as possible. Detailed studies of terrain and road networks showed that a major Soviet move from the Soviet border down through the Zagros Mountains would be difficult and vulnerable.[40]

A more extreme view held that the United States could not expect to deter a Soviet military move into Iran by conventional means. The only effective response would be to threaten the use of nuclear weapons—in effect, threatening global war—or to resort to "horizontal escalation," jargon for counterattacking the Soviets elsewhere.[41] Neither alternative is attractive for the United States in comparison with the policy of developing a credible conventional deterrent in the region itself.

Working to the American advantage was the fact that the greater interest in the Gulf was clearly on the U.S. side, and the Soviets seemed to know this. As a result they might be expected to take fewer risks in the area than would the United States. In addition, the Soviets could never be sure, whatever stated policy might be, that military operations in the Gulf could be isolated regionally, and thus the overall threat of nuclear war would enter their calculations. Last, as time went on the Americans were developing an ability to intervene in the Gulf that could significantly hamper an all-out Soviet military effort.

By creating a force aimed at the Soviet threat, the United States was also becoming able to deal with a whole range of lesser military contingencies. Thus any attempt to close the Strait of Hormuz by a regional power, or any threat to Saudi oil facilities or tankers by Iran or Iraq, could be met by U.S. forces. U.S. forces might also be able to provide timely assistance to a regime that confronted internal problems, but this is the case in which they would probably be least useful. In the face of internal threats, forces from within the region—for example, from Pakistan, Jordan, or Egypt— might be much more acceptable to a threatened regime. This was part of the logic behind U.S. logistical assistance to the Jordanian army.

Thus while the United States has a greater ability to operate militarily in the region than it did ten years ago, the continuing limitations suggest that no policy that relies on American military power in the area will be able to cope with the full range of dangers. Diplomacy, backed by credible force, will still be an essential ingredient of American energy security in the late 1980s.

40. McNaugher, *Arms and Oil*; and Epstein, "Soviet Vulnerabilities."
41. Joshua M. Epstein, "Horizontal Escalation: Sour Notes of a Recurrent Theme," *International Security*, vol. 8 (Winter 1983–84), pp. 19–31 (Brookings Reprint 401).

Achieving a high degree of energy security will require U.S. diplomatic initiatives that take into account the complexity of the issue. First and foremost, energy security is an international problem; despite its comparatively comfortable supply of domestic energy resources, the United States will not be able to isolate itself from the effects of future major oil-supply disruptions. Second, the U.S. reaction to any future energy crisis will have a major impact on all other oil importers.

The purpose of diplomacy as part of energy security is to reduce the likelihood of future oil-supply disruptions, to lay the foundation for cooperation among the industrialized countries in the event of oil shocks, and, in the event of extreme emergencies, to prepare the way for military action to ensure the flow of oil. In brief, diplomacy can help prevent oil crises, can shorten their duration, and can lessen their effects.

The starting point for such diplomacy is to identify the key oil-exporting countries in which American action might enhance stability. Saudi Arabia, Mexico, and Nigeria are all high on this list, along with a second echelon that includes Venezuela, Indonesia, and perhaps Iraq. In each case, some concrete steps by the United States could enhance stability—such as economic and military assistance—and others might be destabilizing, as was proved in the case of Iran in the mid-1970s. There is no simple formula for success, but policymakers should be especially attentive to the need for careful management of these bilateral relations.

A task more complex than the development of bilateral relations confronts the United States: creating a regional environment conducive both to the export of oil and, in emergencies, to U.S. intervention to restore the flow of oil. For this purpose countries such as Egypt, Oman, Jordan, Pakistan, Turkey, and perhaps Israel can play a variety of roles. In addition to the Persian Gulf region, the Red Sea area and the Caribbean become important to U.S. energy security.

The problem in trying to influence the broader regional environment is that there are definite limits on American power in regions like the Middle East. Grand plans aimed at forging a "strategic consensus" have foundered repeatedly on the rocks of local political rivalries. In addition, some of the regimes most dependent on American military power in the event of a major external threat are also the ones that are most reluctant to accept an overt embrace for reasons of domestic politics. For example, American bases in Saudi Arabia may make military sense but be counterproductive when looked at from the perspective of internal political stability; likewise U.S. military intervention may be militarily more efficient but be less sound on political grounds than actions taken by regional powers.

This analysis does little to clarify the hard choices that would have to be made in any future crisis, but it does direct attention to the fact that some countries on the margins of the oil-producing areas, such as Egypt, Jordan, and Turkey, may in fact be important U.S. partners in future oil crises. Thus some portion of the assistance provided to them should be seen as a valid investment in energy security. It will not always be good domestic politics to spell out these linkages publicly to a skeptical Congress, but they should be understood nonetheless.

In addition to developing its ties to regional powers in the Middle East, the United States has an obvious interest in cooperating with its allies in Europe and Japan. The argument is often made that the allies should do more to provide for their own energy security. They, after all, are more heavily dependent on oil imports than the United States. This is the same point made in the field of defense spending, and the answer is essentially the same: the allies should do more, but the United States nonetheless cannot credibly walk away from the problem even if the allies do not increase their efforts. Their vulnerabilities have become ours, and nothing short of a dramatic retreat to isolationism and economic autarky will change that.

The allies have several important contributions to make to energy security other than simply spending more on defense and devising strategies to meet future energy crises. On the political front, there is a potential division of labor that could enhance stability in regions like the Middle East. Britain has an important position in Oman; the French have good relations with Iraq; and the Germans and the Japanese are deeply involved with the Iranian economy. These links may on occasion cause problems within the Western alliance, but on balance they constitute important elements in creating conditions for regional stability and development.

Even with the best efforts at preventive diplomacy, there are bound to be developments in the Middle East, and perhaps in Latin America and Africa as well, that threaten to disrupt oil supplies. Chief among these are internal upheavals, possibly linked to the ups and downs of the oil market or fueled by ideological currents, such as Islamic fundamentalism, hostile to the West. The economic roots of instability may be subject to various economic remedies, but alienation coupled with Islamic extremism is another matter altogether. In the crucial region of the Gulf it poses a distinct threat to major oil-producing countries.

49

Islamic movements in the Middle East are not simply a reflection of the Iranian revolution, although the success of Khomeini in overthrowing the shah had a profound effect elsewhere in the Middle East. Local issues combine with the more universal message of militant Islam to create quite specific conditions in each Islamic country. But the net effect is often the same: to weaken the base of support of incumbent regimes; to challenge close economic and military cooperation with the West; and to oppose a peaceful resolution of the Arab-Israeli conflict.

In time, Islamic extremism, like radical Arab nationalism in the 1950s and 1960s, may run its course. It will not be able to find easy solutions for the multiple social and economic problems facing the Muslim world. Nor does the Islamic current provide much of an outlet for those who seek greater democratization of their political systems. Nonetheless, for some period of time, radical Islam will be a significant force, and there is relatively little the United States can do to blunt its hostility.

Several steps could be taken, however, that would help limit the appeal of militant Islam. First, some degree of tilting toward Iraq in the current Gulf conflict would lower Iran's chances of exporting its revolution to Iraq or to the Arab side of the Gulf. Second, it is important for the West to promote a fair settlement in Lebanon that will address the concerns of the large Shi'ite population there. This would help to dispel the notion that the West is irrevocably hostile to Islamic movements and, if successful, would also help to bring some stability to Lebanon. Third, the Palestinian cause remains an important symbolic issue in the Islamic world, and any significant movement toward a political solution of this long-standing problem would go some distance to defuse the hostility of many Muslims toward the West. Fourth, Egypt's reentry into the Arab world as a moderating influence should be welcomed by the United States even if at times this move may cause strains between Washington and Cairo.

In short, the United States faces a complicated diplomatic agenda as it works for energy security in the years ahead. In addition to the obvious need to concentrate on specific bilateral relations with oil-producing states, there is much to be done with the Western allies and Japan, as well as a number of regional powers. In the end, issues of security in regions like the Persian Gulf will depend on political and military balances, to which the United States can make a contribution both directly and indirectly, and on the handling of related problems such as Lebanon and the broader Arab-Israeli conflict. During the remaining years of the 1980s, while the oil market is likely to remain slack, the West would do well to use this time to lessen the danger that regional issues, especially in the

Middle East, might again be the proximate causes of future oil shocks. To this end, diplomatic solutions to the Iran-Iraq war, to the problems of Lebanon, and to the Arab-Israeli conflict would all be significant contributions to energy security in the 1990s.

Summary

Political conditions, not economics, will be the primary source of instability in world oil markets over the next several years. In this broad sense the energy security problem is unchanged from the last decade. In terms of fundamentals, however, the problem is quite different for the future.

Market incentives will serve to help stabilize the flow of oil and the market price. Moreover, in the event that a temporary interruption of supplies occurs, the market is capable of adjusting with less trauma than in earlier crises. This conclusion does not depend only on the existence of a buffer provided by excess production capacity and government-controlled strategic storage. The market has adapted to the risks of oil shocks and, as noted above, these adaptations can affect energy security policy.

One cannot be as sanguine about political stability. There are many plausible scenarios that could lead to an interruption of oil supplies, though some are less risky than others. For example, the Arab-Israeli conflict and internal unrest in Saudi Arabia, while the focus of concern in the past decade, will not be the most threatening problems for energy security for the next few years. Similarly, localized events involving temporary damage to oil-production facilities or disruption of tanker traffic are not the most dangerous developments on the horizon and will not have a lasting impact on oil prices.

The most serious threat to energy security would be a major shift in the balance of power in the Gulf region and a corresponding shift in political influence over oil production and pricing decisions. Political factors could drastically undermine the economic incentives for stability. Such would be the case, for example, if Iran succeeded in gaining a dominant position in the Gulf region. No additional physical damage to oil-production facilities need occur to yield a long-term structural change in supply conditions.

Shifts in the balance of power, once they occur, are also difficult to counter through diplomatic and economic policies. Military assistance is less effective without a specific focus and could increase rather than moderate tensions in the area. Similarly, energy-security policies involving strategic petroleum reserves and international coordination are not designed to handle long-term structural shifts in oil supply. Oil consumers would be forced to depend even more on higher-cost sources of energy and to adjust consumption to a lower growth path.

Since reactive policies would be largely ineffective in coping with such structural change, the best defense is one that reinforces the existing power balance. This may be helped by building on the commonality of political and economic interests between Western industrialized countries and various Middle Eastern states with the goal of enhancing stability in the critical Gulf region.